MATHS
GCSE
• *with answers* •

Revision for
MATHS
GCSE
with answers

R C SOLOMON

JOHN MURRAY

Revision guides:

Revision for Chemistry	GCSE *with answers*	0 7195 7632 6
Revision for English	Key Stage 3	0 7195 7025 5
Revision for French	GCSE *with answers and cassette*	0 7195 7306 8
Revision for German	GCSE *with answers and cassette*	0 7195 7309 2
Revision for History	GCSE Modern World History	0 7195 7229 0
Revision for Maths	Levels 3–8 Key Stage 3 and Intermediate GCSE *with answers*	0 7195 7083 2
Revision for Maths	GCSE *with answers*	0 7195 7461 7
Revision for Science	Key Stage 3 *with answers*	0 7195 7249 5
Revision for Science	Key Stage 4 *with answers*	0 7195 7422 6
Revision for Spanish	GCSE *with answers and cassette*	0 7195 7394 7

© R C Solomon 1998

First published in 1998
by John Murray (Publishers) Ltd
50 Albemarle Street
London W1X 4BD

All rights reserved. No part of this publication may be reproduced in any material form (including photocopying or storing in any medium by electronic means and whether or not transiently or incidentally to some other use of this publication) without the written permission of the publisher, except in accordance with the provisions of the Copyright, Designs and Patents Act 1988 or under the terms of a licence issued by the Copyright Licensing Agency.

Illustrations by Wearset, Boldon, Tyne and Wear

Layouts by Fiona Webb
Cover design by John Townson/Creation

Typeset in 12/14 pt Times by Wearset, Boldon, Tyne and Wear
Printed and bound in Great Britain by Alden Press, Oxford

A catalogue entry for this title is available from the British Library

ISBN 0 7195 7461 7

Contents

Introduction		vii
Coursework, revision and exam hints		viii

■ Section 1 Number

Chapter 1	**Operations on numbers** The four operations. Factors. Order of operations	2
Chapter 2	**Types of number** Whole numbers. Fractions. Decimals. Negative numbers	7
Chapter 3	**Calculation** Mental arithmetic. Arithmetic on paper. Use of a calculator. Arithmetic of fractions	14
Chapter 4	**Percentages and ratio** Percentages. Finding the percentage. Compound interest and reverse percentages. Ratio	20
Chapter 5	**Powers** Squares, cubes, roots. Indices. Standard form	27
Chapter 6	**Error and approximation** Rounding. Decimal places and significant figures. Checking by approximation. Appropriate levels of accuracy	34
Chapter 7	**Practical arithmetic** Rates and averages. Money matters. Tables and charts	40
Mixed exercise 1		**45**

■ Section 2 Algebra

Chapter 8	**Algebraic expressions** Substitution. Making formulae. Sequences	48
Chapter 9	**Equations** Equations with one unknown. Solution by trial and improvement. Simultaneous equations. Solving problems with equations. Inequalities	54
Chapter 10	**Algebraic manipulation** Expansion and simplification. Factorisation. Changing the subject	61
Chapter 11	**Coordinates and graphs** Coordinates. Interpretation of graphs. Travel graphs	67
Chapter 12	**Functions and graphs** Graphs of functions. Reciprocal and quadratic graphs. Solving equations by graph. Two-dimensional inequalities	74
Mixed exercise 2		**82**

CONTENTS

■ Section 3 Shape and space

Chapter 13   **Plane figures**   86
Angles and lines. Polygons. Circles

Chapter 14   **Measures, lengths and areas**   94
Units and change of units. Length. Area. Combinations of shapes

Chapter 15   **Solids and volumes**   101
Types of solids. Drawing and making solids. Volume. Combinations of solids. Dimensions

Chapter 16   **Construction and maps**   108
Measurement and construction. Maps. Locus

Chapter 17   **Similarity and Pythagoras**   113
Scale diagrams. Congruent and similar triangles. Pythagoras' theorem

Chapter 18   **Trigonometry**   119
Finding the side. Finding the angle

Chapter 19   **Transformations**   124
Translations and reflections. Rotations and enlargements. Centres of rotation and enlargement. Symmetry

Mixed exercise 3   131

■ Section 4 Handling data

Chapter 20   **Collecting data**   134
Reading tables. Questionnaires. Tallies and frequencies

Chapter 21   **Pictures of data**   140
Pictograms, bar charts, pie charts. Construction of pie charts. Bar charts from frequency tables; frequency polygons. Scatter diagrams

Chapter 22   **Analysing data**   147
Averages. Averages from frequency tables. Cumulative frequency

Chapter 23   **Probability**   154
Single probability. Listing the possibilities. Combinations of probabilities

Mixed exercise 4   160

■ Mock examinations

Exam 1 (Foundation)   164

Exam 2 (Foundation)   168

Exam 3 (Intermediate)   172

Exam 4 (Intermediate)   177

■ Solutions   181

■ Glossary   198

■ Index   199

Introduction

■ The aim of this book

This book has been written to help you pass the GCSE examination in Mathematics, at Foundation or Intermediate level. It contains the material you need to revise for the exam, to consolidate your knowledge of the subject, and to practise on questions similar to those of the exam.

■ The structure of this book

This book contains the following:

- 23 chapters, covering the syllabus for GCSE at Foundation and Intermediate levels. The chapters are grouped according to the four programmes of study, i.e. *Number*, *Algebra*, *Shape and Space*, *Handling Data*
- explanations of topics and definitions of terms included in the syllabus
- worked **Examples** covering the topics of the syllabus
- graded **Exercises** to practise on and to test your understanding
- **Key points** to pinpoint what you need to know and errors that you might be tempted to make
- **Revision checklists** at the end of each chapter, for you to ensure that you have revised and understood the material
- **Mixed exercises** at the end of each of the four sections, to test your ability in answering a variety of questions
- **Mock exams** to provide practice before the real thing! It is expected that you use a calculator, unless a question says otherwise. These exams should take about $1\frac{1}{2}$ hours

The topics not required at Foundation level are indicated by a tinted side panel. This is only a rough guide as the syllabus varies between different boards. You may find that a topic that is not in *your* Foundation syllabus is not indicated.

© IT IS ILLEGAL TO PHOTOCOPY THIS PAGE

Coursework, revision and exam hints

You have been working towards GCSE for a long time. If your syllabus includes coursework, your projects count towards the final mark. Make sure that the effort you spend on them is not wasted!

The final few weeks before the exam are vital – make sure you make good use of your revision time.

The time in the exam is the most important of all! Make the most of the precious minutes in the exam room.

■ Coursework hints

- Structure is very important. Divide your project report into sections, and make clear the point of each section.
- Put in an explanation. Don't just state the results you obtained, explain how you got them and why you think they are correct.
- Include your failures as well as your successes. Even the most brilliant mathematician doesn't get the right answer immediately. If you made a guess and later found that it was wrong, put that into your report.
- Your projects are about mathematics. Don't put all your effort into beautiful diagrams or clever computer applications.
- Use algebra. Try to express mathematical results in algebraic symbols, rather than in words.
- Avoid repetition. There's no point in writing down every single result obtained if they are all essentially the same.

■ Revision hints

- Start your revision early. Don't leave it until the week before the exam.
- Don't try to do too much at one time. An hour per day for six days is better than six hours non-stop.
- Avoid distractions. Don't try to combine revision with watching TV, etc.
- Work with pen and paper, so you can write as well as read. Don't try to convince yourself that you can find the answer without writing it down.
- Practise on mock exams and past papers, so that you know what to expect when you enter the exam room.

■ Exam hints

- Read the questions carefully. Many marks are wasted through not understanding what is expected.
- Recognise what each question is about. If a question has several parts, then often the parts follow on from each other.
- Do your best questions first. Try to gain as many marks as possible in the first half of the exam.
- If you can't see how to do a question, leave it and come back to it later. The penny may have dropped!
- Use your time well. Don't spend 20 minutes on a question which is only worth 2 marks. Don't spend ages drawing a beautiful diagram – you are not being tested for your artistic ability.
- Show your working. If you make a slight mistake, but do the rest of the working out correctly, you will get most of the marks. If you write down a wrong answer with no working then you will get no marks.
- If you have time at the end, check that your answers are correct and easy to read. But don't spend too long on this.
- Above all, make good use of your time!

Section 1
NUMBER

CHAPTER 1 Operations on numbers

> **Chapter key points**
> - Read a question carefully to see which number operation is required. For example, if five people each have £300, the total is found by *multiplying* £300 by 5.
> - The order of operations matters – if there are no brackets, make sure that you do the multiplication or division before the addition or subtraction. If there are brackets, do the operation inside the brackets first.

A great amount of mathematics consists of numbers and the things we do to them. We add numbers, subtract them, multiply them and divide them. These are called number **operations**.

1.1 The four operations

The four basic operations are addition, subtraction, multiplication and division.

Addition

The following all mean the same. In each case the answer is 8.

$5 + 3$ the sum of 5 and 3   the total of 5 and 3   5 added to 3

Subtraction

The following all mean the same. In each case the answer is 2.

$6 - 4$ the difference of 6 and 4   6 less 4   4 taken from 6
4 subtracted from 6

Multiplication

The following all mean the same. In each case the answer is 10.

2×5 the product of 2 and 5   2 multiplied by 5   2 times 5

Division

The following all mean the same. In each case the answer is 4.

$8 \div 2$ 8 divided by 2   8 over 2   2 into 8   8/2

A division may not be exact. 3 goes into 7 twice, with 1 left over.

$7 \div 3$ is 2, with 1 left over.

■ Example 1.1

In the morning I drove 56 miles, and in the afternoon I drove 22 miles. What was the total distance?

Solution
Find the sum of 56 and 22. This is 78.
I drove 78 miles.

■ Example 1.2

Judith has seen a compact disc player for £245. She has saved up £67. How much more does she need?

Solution
She needs the difference between £67 and £245, i.e. £178.
She needs £178.

Example 1.3
Certain bottles of wine cost £4.00 each. How many can be bought for £30, and how much money will be left over?

Solution
Divide 30 by 4. The result is 7, with 2 left over.
Seven bottles can be bought, and £2 is left over.

Key point
- Read a question carefully to see which operation is required. For example:
  - If five people each have £300, the total is found by *multiplying* £300 by 5.
  - If £600 is shared between four people, then each share is found by *dividing* £600 by 4.

EXERCISE 1A

1. A car weighs 950 kg, and its driver weighs 70 kg. What is the total weight of car and driver?

2. A pair of shoes costs £33. How much change will there be from two £20 notes?

3. A 25 year mortgage will be paid off in 2010. When was it taken out?

4. A dealer buys a car for £5500 and sells it for £6875. What was the profit?

5. A woman was born in 1948. How old will she be on her birthday in 2006?

6. A cricket team needs to score 278 to win a match. How many more runs are needed after scoring 159?

7. A tennis club sells its used balls in packs of six. How many packs can be made up from 100 balls, and how many balls will be left over?

8. A woman has six grandchildren. If she shares 20 satsumas between them fairly, how many does each child get, and how many satsumas will be left over?

9. Suppose that the EU decimalises the calendar so that a week is 10 days long. How many weeks will there be in a year of 365 days? How many days will be left over at the end?

10. In some fairy stories, the hero wins a pair of 'seven league boots'. Wearing these boots, each stride takes him seven leagues (1 league = 3 miles). If he is to travel 400 miles, how many strides should he take? How many miles will be left to travel after he has taken the boots off?

11. You need 97p for a package. How can you make it up from 17p stamps and 20p stamps?

12. You have two buckets: Bucket A holds 7 pints and Bucket B holds 5 pints. You can fill them from a tap. How can you use the buckets to measure out the following quantities exactly?
    a) 12 pints   b) 15 pints   c) 19 pints

13. Fill in the boxes so that the following are correct.
    a) ☐ + 8 = 21   b) ☐ × 7 = 28
    c) 50 ÷ ☐ = 2   d) 23 − ☐ = 7

OPERATIONS ON NUMBERS

1.2 Factors

The **multiples** of a number are found by multiplying it with other numbers. The multiples of 3 are 3, 6, 9, 12, etc.

The **factors** of a number divide exactly into it. The factors of 12 are 1, 2, 3, 4, 6 and 12.

A **prime** number has no factors apart from 1 and itself. The prime numbers up to 20 are 2, 3, 5, 7, 11, 13, 17 and 19.

A number divisible by 2 is **even**. A number not divisible by 2 is **odd**. 14 is even. 13 is odd.

A number multiplied by itself gives a **square** number. 9 is 3×3, hence 9 is a square number.

Tests for some simple factors are as follows:

- for 2 – a number divisible by 2 ends in 0, 2, 4, 6 or 8.
- for 3 – a number is divisible by 3 if the sum of its digits is divisible by 3.
- for 5 – a number divisible by 5 ends in 0 or 5.

■ Example 1.4

Find the prime factors of the following numbers.

a) 42   b) 17

Solution

a) 42 is divisible by 2.
$42 = 2 \times 21$
The sum of the digits of 21 is $2 + 1$, i.e. 3. Divide 21 by 3. 21 is 3×7.
The prime factors of 42 are 2, 3 and 7.
This could also be solved using a factor tree:

b) The only divisors of 17 are 1 and 17.
It is a prime number.
The only prime factor of 17 is 17.

EXERCISE 1B

1. Which of the following are prime numbers? Find the prime factors of those that are not prime.

   a) 11   b) 22   c) 35   d) 12   e) 25

2. Danielle has 12 squares of cardboard to be arranged in a rectangle. One arrangement is shown in Fig. 1.1. What other arrangements are there?

3. Hen eggs are sold in rectangular boxes of 6, which are arranged 2 eggs by 3 eggs. Smaller eggs, such as quail eggs, are sold in rectangular boxes of 20. What could be the arrangement of these boxes?

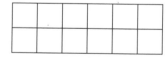

Fig. 1.1

4. In each of the following sentences, insert one of the words *factor, multiple, prime, square, even*.

   a) 24 is a . . . of 8.   b) 49 is a . . . number.
   c) 9 is not a . . . number.   d) 23 is a . . . number.
   e) 5 is a . . . of 35.   f) 33 has no . . . factors.

5. In each of the following sentences, insert one of the numbers 2, 15, 28 or 36.

   a) . . . is a multiple of 7.   b) . . . is a square number.
   c) . . . is an odd number.   d) . . . is a prime number.

6. A number is divisible by 9 if the sum of its digits is divisible by 9. Use this to show that 295 245 is divisible by 9.

> **7** A number is divisible by 6 if it is divisible by 2 and 3. Which of the following are divisible by 6?
>
>   **a)** 435 223 **b)** 672 102 **c)** 661 778 **d)** 445 191
>
> **8** Which of the following statements are true and which are false?
>
>   **a)** The sum of two odd numbers is odd.
>   **b)** The product of two even numbers is even.
>   **c)** The product of an odd number and an even number is even.
>   **d)** The difference of an odd number and an even number is even.

1.3 Order of operations

Suppose you add several numbers. It doesn't matter which numbers you add first.

$$4 + 5 + 6 = 4 + 11 = 15 \quad \text{(adding 5 and 6 first)}$$
$$4 + 5 + 6 = 9 + 6 = 15 \quad \text{(adding 4 and 5 first)}$$

Similarly, if several numbers are multiplied, the order does not matter.

But if an expression involves addition *and* multiplication, then the order does matter. Multiply first.

$$2 + 3 \times 5 = 2 + 15 = 17$$

If you want to add first, put the addition in **brackets**.

$$(2 + 3) \times 5 = 5 \times 5 = 25$$

Note that the answers are different.

In general, *multiply or divide before adding or subtracting*. If you want to add or subtract first, use brackets.

If an expression has subtractions and additions, do the operations from left to right.

■ Example 1.5

Evaluate the following.

a) $24 + 6 \div 2$ **b)** $(24 + 6) \div 2$ **c)** $9 - 3 - 2$ **d)** $9 - (3 - 2)$

Solution

a) Divide first, obtaining $24 + 3$. Then add.
$$24 + 6 \div 2 = 27$$

b) Add first, obtaining $30 \div 2$. Then divide.
$$(24 + 6) \div 2 = 15$$

c) Subtract from left to right. Subtract 3 from 9, obtaining 6. Then subtract 2.
$$9 - 3 - 2 = 4$$

d) First subtract inside the brackets, obtaining 1. Then subtract this from 9.
$$9 - (3 - 2) = 8$$

> **Key point**
>
> ■ The order of operations matters – if there are no brackets, make sure that you do the multiplication or division before the addition or subtraction. If there are brackets, do the operation inside the brackets first.

EXERCISE 1C

1. Evaluate the following.

   a) $7 \times 2 + 3$
   b) $7 \times (2 + 3)$
   c) $12 - 3 \times 4$
   d) $(12 - 3) \times 4$
   e) $24 \div 6 + 2$
   f) $24 \div (6 + 2)$
   g) $14 - 3 + 5$
   h) $14 - (3 + 5)$

2. A book dealer buys 30 books at £1.00 each and 40 books at £2.00 each. How much does he spend in total?

3. When they went on holiday to France, Mr Smith took £200 and Mrs Smith took £300. How many French francs did they get at 9FF per £?

4. What is the total weight of 12 nuts at 2 grams each and 20 bolts at 3 grams each?

5. How much change do I get, if I pay with a £50 note for a £20 suit on which there is a reduction of £2?

6. Cakes cost 20p each at the school fayre. Liam buys eight and Ben buys five. How much do they spend in total?

7. Seven glass jars each contain 2 litres of water. How much water is left after two jars have been emptied?

8. 'Milko' chocolate bars cost 15p each. Yvonne has 62p and Zoe has 58p. How many bars can they buy if they put their money together?

9. A pools win of £12 000 is shared between two men and four women. How much does each person get?

Revision checklist

This chapter has revised:

1.1 The four operations of $+$, $-$, $\times$ and $\div$. ☐
1.2 Factors and multiples. Odd, even, prime and square numbers. ☐
1.3 The order of operations, and the use of brackets. ☐

CHAPTER 2 Types of number

Chapter key points

- Don't leave out zeros in numbers. The number four thousand and twenty-three is 4023, not 423.
- When simplifying a fraction, don't *subtract* the same number from top and bottom:

  $\frac{6}{7} \neq \frac{5}{6}$

- When subtracting a negative number, you add a positive number.
- When you multiply or divide two negative numbers, the result is positive.
- Going to the left on the number line, numbers get smaller. For example, $-5 < -3$.

2.1 Whole numbers

We have ten fingers and thumbs. Hence we use a **base 10**, or **denary**, system of numbers. The value of a digit depends on where it is in a number. In the number

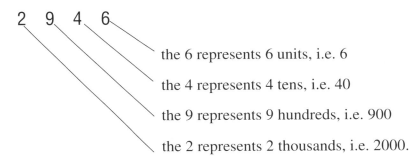

the 6 represents 6 units, i.e. 6
the 4 represents 4 tens, i.e. 40
the 9 represents 9 hundreds, i.e. 900
the 2 represents 2 thousands, i.e. 2000.

■ Example 2.1

A car is reduced from £5895 to £5695. What is the reduction?

Solution

The 8 has been changed to a 6. The 8 and the 6 represent 800 and 600 respectively.
The reduction is £200.

■ Example 2.2

a) Write 4067 in words.
b) Write five million two thousand in figures.

Solution

a) The 4 represents 4000, i.e. four thousand. There are no hundreds. The 6 represents 60, i.e. sixty. The 7 represents seven.
   Four thousand and sixty-seven
b) Five million is 5 000 000. Two thousand is 2000. Combine these.
   5 002 000

Key point

- Don't leave out zeros. The number four thousand and twenty-three is 4023, not 423.

EXERCISE 2A

1. Write down the following numbers in words.

   a) 184 b) 2105 c) 6 200 000 d) 62 320 000

2. Write down the following numbers in figures.

   a) Three thousand four hundred and twenty-seven
   b) Five hundred and thirteen thousand, six hundred and eighteen
   c) Seven hundred thousand and thirty-seven

3. What is the next whole number after:

   a) 3017 b) 89 c) 999?

4. What is the last whole number before:

   a) 318 b) 4000 c) 1 000 000?

5. The population of Japan is about a hundred and twenty million. Write the population in figures.

6. The mileometer of a car has 5 digits, so that after 99 999 miles it starts again at 0. It now registers 95 000. What will it register after 8000 more miles?

7. You can use the digits 6, 7, 5 and 8 to make a four digit number. What is the largest number you can make? What is the smallest?

8. Write 320 in words. Multiply this number by 100, giving your answer both in figures and in words.

9. Add two hundred and twenty-three to three hundred and forty-seven, giving your answer in words.

10. The ancient Babylonians used the symbol | for 1 and the symbol ⟨ for 10.

    Hence 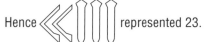 represented 23.

    a) Convert 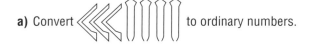 to ordinary numbers.

    b) How would the Babylonians write 47?

11. In the number 435, what does the 3 represent? Multiply 435 by 100. What does the 3 now represent?

2.2 Fractions

The top number of a fraction is the **numerator**. The bottom number is the **denominator**. In the fraction $\frac{3}{8}$, the numerator is 3 and the denominator is 8.

A fraction can be simplified by dividing top and bottom by the same number:

$$\frac{6}{16} = \frac{3}{8}$$

In a **proper** fraction, the numerator is less than the denominator. In an **improper** fraction, the numerator is greater than the denominator.

$\frac{5}{8}$ is proper $\frac{18}{13}$ is improper

A **mixed number** contains a whole number and a fraction. $2\frac{3}{4}$ is a mixed number.

Example 2.3

a) Simplify the fraction $\frac{21}{28}$.
b) Convert $\frac{14}{3}$ to a mixed number.
c) Convert $3\frac{2}{5}$ to an improper fraction.

Solution

a) Divide the numerator and the denominator by 7.
$\frac{21}{28} = \frac{3}{4}$

b) Divide the numerator by the denominator. 3 goes into 14 four times, and there is 2 left over.
$\frac{14}{3} = 4\frac{2}{3}$

c) The whole number 3 is equal to $\frac{15}{5}$. Add this to $\frac{2}{5}$.
$3\frac{2}{5} = \frac{17}{5}$

Key point

- When simplifying a fraction, don't *subtract* the same number from top and bottom.
$\frac{6}{7} \neq \frac{5}{6}$

EXERCISE 2B

1. Write as fractions:
    a) two thirds
    b) a tenth
    c) seven tenths

2. Write in words:
    a) $\frac{1}{2}$
    b) $\frac{2}{5}$
    c) $\frac{4}{9}$

3. Simplify the following fractions.
    a) $\frac{2}{4}$
    b) $\frac{8}{12}$
    c) $\frac{15}{20}$

4. Convert the following improper fractions to mixed numbers.
    a) $\frac{7}{5}$
    b) $\frac{11}{7}$
    c) $\frac{12}{5}$

5. Convert the following mixed numbers to improper fractions.
    a) $2\frac{1}{2}$
    b) $1\frac{1}{8}$
    c) $3\frac{2}{3}$

6. Below are some half-finished problems on simplifying fractions. Complete them.
    a) $\frac{3}{*} = \frac{1}{4}$
    b) $\frac{*}{15} = \frac{2}{3}$
    c) $\frac{20}{25} = \frac{*}{5}$

7. From a sample of 500 voters, 125 said they voted Conservative. What fraction voted Conservative?

8. A firm employs 800 people, of whom one fifth are trainees. How many are not trainees?

9. Two cans of cat food are shared fairly between eight cats. What fraction of a tin does each cat get?

10. Fig. 2.1 shows three shapes in which part has been coloured. In each case find the fraction that has been coloured.

    a)
    b)
    c)

Fig. 2.1

TYPES OF NUMBER

> **11 a)** How many seconds are there in half a minute?
> **b)** What fraction of a minute is 15 seconds?
>
> **12** A computer is reduced from £1000 to £800. By what fraction is it reduced?
>
> **13** In a sale, all prices are reduced by a third. What is the reduced price of a suit that originally cost £120?
>
> **14** Find a quarter of a hundred thousand, giving your answer:
> **a)** in figures **b)** in words

2.3 Decimals

A **decimal** is a fraction in which the denominator is 10, 100, 1000, and so on.

$$0.3 = \tfrac{3}{10} \qquad 0.26 = \tfrac{26}{100}$$

The digit after the decimal point is in the **first decimal place**. The digit after that is in the second decimal place, and so on. So in the number

$$0.712$$

7 is in the first decimal place

1 is in the second decimal place

2 is in the third decimal place

Conversion

■ Decimal → fraction

Express the decimal as a fraction with denominator 10, 100, and so on. Simplify it if possible.

■ Fraction → decimal

Divide the numerator by the denominator. If the division stops, the result is a **terminating** decimal. If the division does not stop, it is a **recurring** decimal.

$\tfrac{5}{8} = 0.625$ This is a terminating decimal
$\tfrac{2}{3} = 0.666...$ This is a recurring decimal: the 6s go on forever

Rounding

You can round a recurring decimal correct to 3 decimal places. Look at the digit in the fourth decimal place. If it is 5 or greater, round up. If it is 4 or less, round down.

$\tfrac{2}{3} = 0.6666... = 0.667$ (correct to 3 decimal places)
$\tfrac{1}{3} = 0.3333... = 0.333$ (correct to 3 decimal places)

■ Example 2.4

a) Convert 0.35 to a fraction.
b) Convert $\tfrac{2}{11}$ to a decimal, giving your answer correct to 3 decimal places.

Solution

a) Write 0.35 as $\tfrac{35}{100}$. Divide top and bottom by 5.
$0.35 = \tfrac{7}{20}$

b) Divide 11 into 2. This gives 0.181818... . Round correct to 3 decimal places. The digit in the fourth decimal place is 8, so round up.
$\frac{2}{11} = 0.182$

■ Example 2.5

Find the largest of $\frac{1}{3}$, 0.3 and $\frac{2}{7}$.

Solution

Write all three as decimals.

$\frac{1}{3} = 0.333...$ 0.3 $\frac{2}{7} = 0.2857...$

The largest is $\frac{1}{3}$.

EXERCISE 2C

1. For the number 0.46531, what are the digits:
   a) in the second decimal place
   b) in the fourth decimal place?

2. Convert the following decimals to fractions. Simplify if possible.
   a) 0.7   b) 0.25   c) 0.29   d) 0.125   e) 0.68

3. Convert the following fractions to decimals.
   a) $\frac{1}{4}$ b) $\frac{3}{5}$ c) $\frac{13}{20}$ d) $\frac{7}{8}$ e) $\frac{3}{1000}$

4. Convert the following fractions to decimals, giving your answers correct to 3 decimal places.
   a) $\frac{1}{3}$ b) $\frac{5}{9}$ c) $\frac{2}{7}$ d) $\frac{5}{13}$ e) $\frac{1}{6}$

5. Convert the following decimals to fractions, giving your answers both as mixed numbers and as improper fractions.
   a) 1.2   b) 2.75   c) 3.85   d) 4.24   e) 6.875

6. Write down the larger of $2\frac{1}{4}$ and 2.3.

7. Write down the smaller of 3.12 and $3\frac{1}{8}$.

8. Arrange the numbers below in increasing order.
   2.3 $2\frac{1}{3}$ 2.29 $2\frac{2}{7}$

2.4 Negative numbers

A **number line** (Fig. 2.2) shows numbers in order:

Fig. 2.2

Positive numbers are to the right of 0. **Negative** numbers are to the left of 0.

To show that one number is greater than another, use the > or < symbols. So, for example:

$3 < 7$ $-5 < -1$ $6 > -2$

Subtracting a negative number is the same as adding a positive number.

$12 - (-7) = 12 + 7 = 19$

The product or quotient of negative numbers is positive.

$(-3) \times (-5) = 15$ $(-20) \div (-5) = 4$

TYPES OF NUMBER

■ Example 2.6

The thermometer in Fig. 2.3 is marked in degrees Celsius (°C). What is the difference in temperature between 10 °C and −8 °C?

Solution
Subtract −8 from 10.
$10 - (-8) = 10 + 8 = 18$
The difference in temperature is 18 degrees.

Fig. 2.3

Key points

The following exercise is about the arithmetic of negative numbers. Don't forget:

- When subtracting a negative number, you add a positive number.
- When you multiply or divide two negative numbers, the result is positive.
- Going to the left on the number line, numbers get smaller. For example, $-5 < -3$.

EXERCISE 2D

1 Evaluate the following.

a) $(-4) \times (-6)$ b) $(-2) \times (-2) \times (-2)$ c) $(1-3) \times (2-6)$
d) $8 \div (-2)$ e) $(-9) \div 3$ f) $(-12) \div (-4)$

2 Fig. 2.4 shows a number line. Mark on it:

a) -1 b) 1.5 c) -0.5 d) $(-1) \times (-1)$

Fig. 2.4

3 A cliff is 100 m high. A gannet is flying 200 m above the top of the cliff, then dives to the sea below. How far does it dive?

4 A ship starts 250 miles east of the Greenwich meridian. It sails to 380 miles west of the meridian. How far west has it sailed?

5 Sarabjit's bank statement says −£205, i.e. she has an overdraft of £205. Last week she had £315 in her bank account. How much has gone from her account over the week?

6 The table below gives the names of some cities from around the world, and the number of hours their local time is ahead of Greenwich Mean Time (GMT).

City	Tel Aviv	New York	Hong Kong	Vancouver
Hours ahead	2	−5	8	−8

a) What is the time difference between Tel Aviv and New York?
b) When you travel from Vancouver to New York, by how much do you need to alter your watch?

7 The level of a reservoir is measured over several years. The table below gives the average height in metres above the original level.

Year	0	1	2	3	4	5	6	7
Level (m)	0	1.5	2.3	0.2	−1.7	−2.9	−0.1	0.8

a) What was the lowest level recorded?
b) Between which years was there the greatest rise in level?

8 List the following in increasing order, using the < symbol.

$$-4 \quad 4 \quad -1 \quad 0 \quad -3 \quad 5 \quad -6$$

9 List the following in decreasing order, using the > symbol.

$$\tfrac{1}{2} \quad -0.1 \quad -\tfrac{1}{3} \quad -1 \quad 1 \quad \tfrac{1}{4}$$

10 During the French Revolution, the calendar was reformed by counting years after 1789, the fall of the Bastille prison.

a) What is the AD date corresponding to the French Revolutionaries' Year 6? What corresponds to Year −15?
b) What date would the French Revolutionaries give to AD1805? What to AD1688?

Revision checklist

This chapter has revised:

2.1 Writing numbers in words and in figures. ❑
2.2 Simplifying fractions, and converting between mixed numbers and improper fractions. ❑
2.3 Writing fractions as decimals, and converting decimals to fractions. ❑
2.4 Showing negative numbers on a number line, and the arithmetic of negative numbers. ❑

CHAPTER 3 *Calculation*

Chapter key points
- If an exam question asks you to evaluate an expression without a calculator, then show all your working. Don't rub anything out. Of course, you can use your calculator to *check* your answer.
- Different calculators require different sequences of keys. Make sure you know the sequences for your calculator.
- Don't forget to press the $=$ key at the end of a calculation.
- A calculator uses the same rules as we do for the order of operations. Be sure to put in brackets when needed, or to press the $=$ key between operations.
- When adding fractions, don't add the numerators together and the denominators together:

  $\frac{3}{4} + \frac{2}{5} \neq \frac{5}{9}$

- When multiplying a fraction by a whole number, multiply only the numerator by the number:

  $\frac{1}{8} \times 3 = \frac{3}{8}$ *not* $\frac{3}{24}$

Normally you use a calculator for calculation. This chapter covers the use of a calculator. It also covers calculation on paper and the arithmetic of fractions.

3.1 Mental arithmetic

For numbers up to 10, you should be able to do arithmetic mentally. When you multiply a whole number by 10, add a 0 to its end. If a number ends in 0, when you divide by 10 remove the 0.

$6000 \times 400 = 2\,400\,000$
$8000 \div 20 = 400$

■ Example 3.1
A plane journey is 3000 miles. How long will it take at 500 m.p.h.?

Solution
Divide 3000 by 500. Cancel two 0s from both terms.
$3000 \div 500 = 30 \div 5$
The journey will take 6 hours.

EXERCISE 3A
Do not use a calculator for these questions.

1. Evaluate:

   a) 5000×300 b) 200×3000 c) $20 \times 70\,000$
   d) $800\,000 \div 40$ e) $5000 \div 20$ f) $4000 \div 2000$

2. What is the total weight of 200 cars at 800 kg each?

3. John is training for a race. Every day he runs 20 miles. Over 40 days, what is the total distance he runs?

4. A lottery win of £12 000 000 is shared between 20 people. How much does each person get?

5. A library has 800 000 mm of shelving. How many books can be stored, if on average a book is 20 mm wide?

3.2 Arithmetic on paper

Arithmetic of numbers with more than one digit can be done on paper. The following gives examples of the four operations.

■ Example 3.2

Use paper to evaluate:

a) $43.2 + 68.2$ **b)** $84 - 27$ **c)** 71×23 **d)** $544 \div 17$

Solution

a) Make sure that the decimal points are aligned. Note the 'carry' when 3 is added to 8, and when 4 is added to 6.

```
  1 1
  4 3 . 2
  6 8 . 2
1 1 1 . 4
```

b) Set the 27 below the 84. Note the 'borrow' when 7 is subtracted from 4.

```
  7 1
  8 4
  2 7
  5 7
```

c) Multiply 71 by 3, obtaining 213. Multiply 71 by 20, obtaining 1420 (don't forget the 0). Add the results.

```
    7 1
    2 3
  2 1 3
1 4 2 0
1 6 3 3
```

d) Divide 17 into 54, then bring down the next 4 beside the remainder, and divide again.

```
        3 2
   17)5 4 4
      5 1
        3 4
        3 4
          0
```

Key point

■ If an exam question asks you to evaluate an expression without a calculator, then show all your working. Don't rub anything out. Of course, you can use your calculator to *check* your answer.

EXERCISE 3B

Evaluate the following on paper, without a calculator.

1. **a)** $451 + 817$ **b)** $23.1 + 89.2$ **c)** $431 - 79$
   d) $74.2 - 24.9$ **e)** 63×29 **f)** 213×17
   g) $1025 \div 41$ **h)** $289 \div 17$

2. Find the total weight of a van of 2875 kg and its load of 617 kg.

3. Silas started the day with £32.27, then spent £9.62. How much did he have left?

4. A booklet of 57 pages has an average of 312 words per page. How many words are there in the whole booklet?

5. A total of 1200 minutes on the Internet has been allocated to a class of 24 students. How long will each student get?

3.3 Use of a calculator

All calculators can perform the four operations of $+$, $-$, $\times$ and $\div$. A scientific calculator can perform many other functions. The instructions below may not work exactly for your calculator.

To evaluate 9×7 press the following:

[9] [×] [7] [=]   **63** appears

More than one operation

A scientific calculator can evaluate a complicated expression in one go. With an ordinary calculator, you may have to do it in stages. For $8 \times 9 + 3 \times 7$:

Scientific calculator

[8] [×] [9] [+] [3] [×] [7] [=]   **93** appears

Ordinary calculator

[8] [×] [9] [=]   **72** appears

[3] [×] [7] [=]   **21** appears

[7] [2] [+] [2] [1] [=]   **93** appears

Brackets

With a scientific calculator you can evaluate expressions in brackets. With an ordinary calculator you must do the calculation in two stages. For $(5 + 7) \times 3$:

Scientific calculator

[(] [5] [+] [7] [)] [×] [3] [=]   **36** appears

Ordinary calculator

[5] [+] [7] [=]   **12** appears

[1] [2] [×] [3] [=]   **36** appears

Memory

Most calculators have a memory key. A scientific calculator may have more than one memory.

The memory is useful for:

- storing the result of a calculation
- evaluating a complicated expression
- storing a number that is used many times in a calculation.

Inverse key

Many keys on a calculator can do more than one task. Often the second task is done by pressing the inverse key first (or second function, or shift). You may have a key on your calculator like the one in Fig. 3.1. It can find squares *and* square roots.

[9] [√]   gives $\sqrt{9}$, i.e. **3**

[9] [INV] [√]   gives 9^2, i.e. **81**

Fig. 3.1

CALCULATION

+/− key

The +/− key makes a positive number negative, and a negative number positive. To evaluate $(-21)^2$, press the following:

[2] [1] [±] [INV] [√]   gives **441**

There are many other keys on a scientific calculator, whose use will be explained in later chapters.

■ Example 3.3

Susan has to evaluate the expression

$$\frac{2.1 + 3.6}{1.2}.$$

She uses her calculator by pressing keys in the sequence below. Explain why the sequence is wrong, and give a correct sequence.

[2] [.] [1] [+] [3] [.] [6] [÷] [1] [.] [2] [=]

Solution

The calculator does the division first. The 1.2 should divide the whole of the top line, not just the 3.6. By Susan's sequence, 1.2 will divide 3.6 but not 2.1.

She should either put the top line in brackets, or press the = key after the 3.6. Two possible sequences are:

[(] [2] [.] [1] [+] [3] [.] [6] [)] [÷] [1] [.] [2] [=]

or

[2] [.] [1] [+] [3] [.] [6] [=] [÷] [1] [.] [2] [=]

In either sequence the solution 4.75 should appear.

Key points
- Different calculators require different sequences of keys. Make sure you know the sequences for your calculator.
- Don't forget to press the = key at the end of a calculation.
- A calculator uses the same rules as we do for the order of operations. Be sure to put in brackets when needed, or to press the = key between operations.

EXERCISE 3C

1. Evaluate the following.

   a) $2.7 \times 3.1 + 3.8 \times 1.9$
   b) $17 \times (56.1 - 32.9)$
   c) $27 \div (44 - 26)$
   d) $\frac{3.121 + 4.49}{1.18}$
   e) $\frac{40.56}{5.19 - 3.24}$
   f) $\frac{27 \times 56}{1.23 + 2.97}$
   g) $(2.4 + 3.2)^2$
   h) $(-556) \times (-5.8)$
   i) $\sqrt{(25^2 - 7^2)}$

2. The calculator sequence below is used to evaluate $\frac{1.23}{3.5 + 2.7}$.

   Explain what is wrong with the sequence and give a correct sequence.

   [1] [.] [2] [3] [÷] [3] [.] [5] [+] [2] [.] [7] [=]

CALCULATION

3 The calculator sequence below is used to evaluate $\dfrac{6.7 + 4.5}{2.8 \times 9.3}$.

Explain what is wrong with the sequence and give a correct sequence.

$\boxed{(}\boxed{6}\boxed{.}\boxed{7}\boxed{+}\boxed{4}\boxed{.}\boxed{5}\boxed{)}\boxed{\div}\boxed{2}\boxed{.}\boxed{8}\boxed{\times}\boxed{9}\boxed{.}\boxed{3}\boxed{=}$

4 The calculator sequence below is pressed. Which of the expressions a), b), c) or d) does it evaluate? Give calculator sequences for the other expressions.

$\boxed{2}\boxed{\div}\boxed{3}\boxed{+}\boxed{5}\boxed{\div}\boxed{4}\boxed{=}$

a) $\dfrac{2+5}{3+4}$ b) $\dfrac{2}{3} + \dfrac{5}{4}$ c) $2 + \dfrac{5}{3+4}$ d) $\dfrac{2+5}{3} + 4$

3.4 Arithmetic of fractions

When multiplying fractions, multiply the tops together and multiply the bottoms together.

To divide by a fraction a/b, multiply by b/a.

When multiplying or dividing mixed numbers, first convert them to improper fractions.

When adding or subtracting fractions, first put them over the same denominator.

■ Example 3.4

Evaluate:

a) $1\tfrac{3}{4} \times \tfrac{2}{11}$ **b)** $\tfrac{2}{7} \div \tfrac{3}{4}$

Solution

a) Convert $1\tfrac{3}{4}$ to the improper fraction $\tfrac{7}{4}$. Multiply the tops of $\tfrac{7}{4}$ and $\tfrac{2}{11}$ together and the bottoms together, obtaining $\tfrac{14}{44}$. Simplify your answer to $\tfrac{7}{22}$.

$\tfrac{7}{4} \times \tfrac{2}{11} = \tfrac{14}{44} = \tfrac{7}{22}$

$1\tfrac{3}{4} \times \tfrac{2}{11} = \tfrac{7}{22}$

b) Dividing by $\tfrac{3}{4}$ is the same as multiplying by $\tfrac{4}{3}$.

$\tfrac{2}{7} \div \tfrac{3}{4} = \tfrac{2}{7} \times \tfrac{4}{3} = \tfrac{8}{21}$

$\tfrac{2}{7} \div \tfrac{3}{4} = \tfrac{8}{21}$

■ Example 3.5

In his will, a man leaves his money to his wife, his son and his daughter. His wife gets $\tfrac{2}{3}$ of the estate, and his son gets $\tfrac{1}{8}$. What fraction of the estate does the daughter receive?

Solution

Add the fractions received by the wife and the son. First write the fractions with a denominator of 24.

$\tfrac{2}{3} + \tfrac{1}{8} = \tfrac{16}{24} + \tfrac{3}{24} = \tfrac{19}{24}$

Now subtract $\tfrac{19}{24}$ from 1, obtaining $\tfrac{5}{24}$.

The daughter receives $\tfrac{5}{24}$ of the estate.

Key points

■ When adding fractions, don't add the numerators together and the denominators together.

$\tfrac{3}{4} + \tfrac{2}{5} \neq \tfrac{5}{9}$

■ When multiplying a fraction by a whole number, multiply only the numerator by the number.

$\tfrac{1}{8} \times 3 = \tfrac{3}{8}$ not $\tfrac{3}{24}$

CALCULATION

EXERCISE 3D

1 Evaluate the following, leaving your answers as fractions.

a) $\frac{1}{3} \times \frac{2}{5}$
b) $\frac{4}{7} \times \frac{3}{11}$
c) $\frac{4}{3} \times \frac{2}{5}$
d) $\frac{3}{10} \times \frac{3}{4}$
e) $\frac{1}{4} \times 12$
f) $\frac{3}{8} \div \frac{2}{3}$
g) $\frac{3}{7} \div \frac{3}{4}$
h) $\frac{3}{10} \div 4$
i) $\frac{4}{9} \div \frac{1}{27}$
j) $1\frac{1}{3} \times 2\frac{1}{2}$
k) $4\frac{1}{5} \times 1\frac{3}{4}$
l) $2\frac{2}{3} \div 3\frac{1}{8}$

2 Evaluate the following, leaving your answers as fractions.

a) $\frac{1}{3} + \frac{3}{8}$
b) $\frac{2}{7} + \frac{1}{9}$
c) $\frac{2}{9} + \frac{1}{4}$
d) $\frac{2}{3} - \frac{1}{5}$
e) $\frac{1}{3} + \frac{1}{9}$
f) $1\frac{1}{3} + 2\frac{1}{4}$
g) $4\frac{1}{5} - 3\frac{2}{3}$
h) $2\frac{7}{8} - 3\frac{1}{4}$
i) $6\frac{1}{10} - 4\frac{1}{4}$

3 A woman wins the National Lottery. She gives $\frac{1}{4}$ of the sum to her brother. Her brother then gives $\frac{1}{3}$ of this amount to his son. What fraction of the original win does the son receive?

4 I buy $\frac{1}{8}$ of a tonne of apples, of which $\frac{1}{5}$ are bad. What is the weight of the bad apples?

5 The Indian rupee is divided either into 100 paisas or into 16 annas. What fraction of an anna is a paisa?

6 If I walk at $3\frac{1}{2}$ m.p.h., how long will it take me to cover 14 miles?

7 A shop is run by four brothers, who share the profits equally. Each brother has to pay $\frac{3}{10}$ of his profit in tax. What fraction of the original profit does each brother retain?

8 In a swimming race, Adele was $\frac{1}{4}$ of a metre ahead of Barbara, and Barbara was $\frac{1}{3}$ of a metre ahead of Charlotte. By how much was Adele ahead of Charlotte?

9 From $2\frac{3}{4}$ m of curtain material $1\frac{2}{3}$ m is cut off. How much material remains?

10 A pint is $\frac{1}{8}$ of a gallon. How much is left after one pint is poured away from a jug which contains $\frac{1}{3}$ of a gallon?

Fractions with a calculator

A scientific calculator may have a key for entering and evaluating fractions. Often it is labelled $a^b/_c$. To enter $\frac{5}{8}$, press the following:

[5] [$a^b/_c$] [8]

To enter $2\frac{3}{4}$, press the following:

[2] [$a^b/_c$] [3] [$a^b/_c$] [4]

The key can convert between mixed numbers and improper fractions. It can also convert a fraction to a decimal.

Practise the use of the $a^b/_c$ button on some of Exercise 3D.

Revision checklist

This chapter has revised:

3.1 Mental arithmetic of whole numbers, multiplying and dividing by 20, 400, etc. ☐

3.2 Arithmetic done on paper. ☐

3.3 Use of a calculator, including brackets, memory, inverse and +/− keys. ☐

3.4 The arithmetic of fractions. ☐

CHAPTER 4 Percentages and ratio

Chapter key points

- When converting percentages, divide by 100, not by 10. Thus 2% is 0.02, not 0.2. Similarly, 0.4 is 40%, not 4%.
- The percentage change of a quantity is a percentage of the *original* quantity, not of the changed quantity. If an item is bought for £40 and sold for £50, the percentage profit is 25%, not 20%.
- Be careful when doing reverse percentages. Suppose an item is sold for £600, at a profit of 20%. The original price is £600 ÷ 1.2 = £500. Don't take 20% off the *selling* price.
- Make sure you get ratios the right way round. If the ratio of A to B is 2 : 3, then A is $\frac{2}{3}$ of B, not the other way round.
- When a quantity is divided in the ratio 2 : 5, the first part is $\frac{2}{7}$ of the whole, not $\frac{2}{5}$.

We often express fractions as percentages and ratios.

4.1 Percentages

A **percentage** is a fraction in which the denominator is 100.
$11\% = \frac{11}{100}$

■ Example 4.1

A bottle of gin is 40% alcohol. How much alcohol is there if the bottle contains 750 ml?

Solution
To find 40% of 750, multiply by $\frac{40}{100}$.
$750 \times \frac{40}{100} = 300$
The bottle contains 300 ml of alcohol.

■ Example 4.2

A computer costs £980 without VAT. What will it cost once VAT at 17.5% is added?

Solution
Find 17.5% of 980. Multiply by $\frac{17.5}{100}$.
$\frac{17.5}{100} \times 980 = 171.5$
Add this on to 980, obtaining 1151.5.
The cost including VAT is £1151.50.

EXERCISE 4A

1. A literary agent takes 10% of her client's earnings. If the client earns £35 000, how much does the agent get?

2. A garden shed costs £680, before VAT at $17\frac{1}{2}\%$ is added. How much is the VAT?

3. In a population of 5 000 000 people, 20% are under 15. How many are under 15?

4. A secondhand car is offered at £6000. The dealer takes 15% off the price. How much is taken off?

5 A brand of sherry contains 18% alcohol. How much alcohol is there in 20 litres of the sherry?

6 The weight of a dog increases by 20%. If it weighed 16 kg before the increase, what does it weigh after the increase?

7 The wages bill for a company is £300 000. If all wages are increased by 6%, what will the new bill be?

8 Derek weighed 18 stone. After going on a diet, his weight has decreased by 20%. How much does he weigh now?

9 A chair is priced at £150. During a sale all prices are reduced by 30%. How much does the chair cost now?

10 Iris wants to tile part of her bathroom. She buys tiles costing £38 and adhesive costing £6. VAT at $17\frac{1}{2}$% is added to the bill. How much does she have to pay in all?

Conversion

■ Decimal → percentage

Move the decimal point two places to the right, i.e. multiply by 100.

$$0.53 = 53\%$$

■ Percentage → decimal

Move the decimal point two places to the left, i.e. divide by 100.

$$3\% = 0.03$$

■ Fraction → percentage

Multiply the fraction by 100.

$$\tfrac{4}{5} \times 100 = \tfrac{400}{5} = 80\%$$

■ Percentage → fraction

Divide by 100, then simplify.

$$25\% = \tfrac{25}{100} = \tfrac{1}{4}$$

Below is a table giving some common conversions.

Decimal	Percentage	Fraction
0.5	50%	$\tfrac{1}{2}$
0.25	25%	$\tfrac{1}{4}$
0.75	75%	$\tfrac{3}{4}$
0.333...	$33\tfrac{1}{3}$%	$\tfrac{1}{3}$
0.666...	$66\tfrac{2}{3}$%	$\tfrac{2}{3}$

Key point

■ When converting percentages, divide by 100, not by 10. Thus 2% is 0.02, not 0.2. Similarly, 0.4 is 40%, not 4%.

PERCENTAGES AND RATIO

EXERCISE 4B

1. Convert the following percentages to fractions in their simplest form.
   a) 30% b) 25% c) 80% d) 2%

2. Express the following fractions as percentages.
   a) $\frac{1}{4}$ b) $\frac{3}{4}$ c) $\frac{1}{20}$ d) $\frac{2}{5}$

3. Express the following decimals as percentages.
   a) 0.43 b) 0.82 c) 0.02

4. Express the following percentages as decimals.
   a) 40% b) 65% c) 4%

5. Arrange the following in increasing order.
   30% $\frac{1}{4}$ 0.2 $\frac{1}{3}$ $\frac{2}{5}$

4.2 Finding the percentage

To express one quantity as a percentage of another, express it as a fraction then multiply by 100.

■ Example 4.3

A school contains 300 boys and 500 girls. What percentage are girls?

Solution

There are 800 pupils in total. The fraction of girls is $\frac{500}{800}$, i.e. $\frac{5}{8}$. Multiply by 100, to obtain 62.5.
The percentage of girls is 62.5%.

■ Example 4.4

The value of a company increases from £120 000 000 to £150 000 000. What is the percentage increase?

Solution

The increase is £30 000 000. Divide this by the original value, then multiply by 100.

$$\frac{30\ 000\ 000}{120\ 000\ 000} \times 100 = 25$$

The increase is 25%.

Key point

■ The percentage change of a quantity is a percentage of the *original* quantity, not of the changed quantity. If an item is bought for £40 and sold for £50, the percentage profit is 25%, not 20%.

EXERCISE 4C

1. In a driving test centre, from 800 candidates 300 passed. What was the percentage pass rate?

2. A tennis club contains 95 men and 65 women. What percentage are women?

3. Anita's salary is £15 000. Her annual rent is £6000. What percentage of her salary goes in rent?

4. The price of a flat increases from £50 000 to £60 000. What is the percentage increase?

5. The weight of a baby increases from 16 lb to 18 lb. What is the percentage increase?

6. The price of a book is cut from £12.00 to £9.00. What is the percentage cut?

7. During a heat wave, the level of water in a reservoir fell from 10 m to 8 m. What was the percentage fall?

4.3 Compound interest and reverse percentages

Suppose money is invested at 10% **compound interest**. This means that every year the money is multiplied by a factor of $\frac{110}{100}$, i.e. of 1.1.

Suppose you are told that after an increase of 10%, a sum of money is £1320. To find the original sum *divide* by 1.1, to obtain £1200.

■ Example 4.5

£5000 is left in a bank at 6% compound interest. The interest is allowed to accumulate. How much is there after three years?

Solution

When a sum of money is increased by 6%, every £100 is increased to £106. Hence the sum of money is multiplied by a factor of 106/100 = 1.06.
After 1 year £5000 is multiplied by 1.06, to become £5300.
After 2 years £5300 is multiplied by 1.06, to become £5618.
After 3 years £5618 is multiplied by 1.06, to become £5955.08.
So after three years there is £5955.08 in the bank.

■ Example 4.6

A house is sold for £58 300, making a profit of 6%. What was the house bought for?

Solution

In the previous example it was shown that adding 6% was equivalent to multiplying by 1.06.
Hence the buying price is given by £58 300 *divided* by 1.06. This gives £55 000.
The buying price was £55 000.

> **Key point**
> ■ Be careful when doing reverse percentages. Suppose an item is sold at £600, for a profit of 20%. The original price is £600 ÷ 1.2, i.e. £500. Don't take 20% off the *selling* price.

EXERCISE 4D

1. The value of a company is increasing at 10% each year. It is valued at £10 000 000 when it starts. What is its value after three years?

2. A saver invests £1000 at 5% compound interest. How much will the savings amount to at the end of three years?

3 A car was bought for £9000. Every year it depreciates in value at 15%. How much is it worth after three years?

4 A radioactive material decays at a rate of 10% by mass each year. Initially there was 2 kg. How much will there be after three years?

5 After a 10% pay rise, a man is earning £19 800. How much did he earn before the rise?

6 A pensioner invests money at 8%. After one year it is £1620. What was the original sum?

7 An antique dealer sells a table for £600, making a profit of 20%. How much was the table bought for?

8 A picture dealer buys a picture, but has to sell it at a loss of 15%. If she sold it for £765, how much did she pay for it?

9 A clothes shop holds a sale in which all prices are reduced by 20%. In the sale, the price of a dress is £64. What was the cost before the reduction?

10 The cost of building works for a house extension was £5170, including VAT at $17\frac{1}{2}$%. What was the cost before the VAT?

4.4 Ratio

If a vinaigrette sauce contains 2 fluid ounces of vinegar for every 5 fluid ounces of oil, then the vinegar and oil are in the **ratio** 2 : 5. The amount of oil is then $\frac{5}{2}$ of the amount of vinegar. The amount of vinegar is $\frac{2}{5}$ of the amount of oil.

Ratios are like fractions. They can be simplified by multiplying or dividing all the terms by the same number.

■ Example 4.7

A firm employs 300 men and 450 women. Find the ratio of men to women, simplifying your answer.

Solution
The ratio is 300 : 450. Both numbers can be divided by 150.
The ratio is 2 : 3.

■ Example 4.8

The ages of Nicholas and Oscar are in the ratio 5 : 6. If Oscar is 36, how old is Nicholas?

Solution
Multiply 36 by $\frac{5}{6}$, obtaining 30.
Nicholas is 30.

Key point
■ Make sure you get ratios the right way round. If the ratio of A to B is 2 : 3, then A is $\frac{2}{3}$ of B, not the other way round.

EXERCISE 4E

1 A school orchestra has 15 boys and 18 girls. What is the ratio of boys to girls?

2 In two towns, the average prices of houses are £60 000 and £75 000 respectively. What is the ratio between these prices?

PERCENTAGES AND RATIO

3. A recipe for a salad dressing specifies 5 parts lemon juice to 20 parts oil. What is the ratio of juice to oil?

4. A woman is five times as old as her daughter. What is the ratio between their ages?

5. Two thirds of the employees of a factory are men. What is the ratio of men to women?

6. For a type of pastry, the ratio of flour to butter is 2 : 1. How much flour goes with 4 ounces of butter?

7. The heights of Hassan and Hussein are in the ratio 8 : 9. If Hassan is 160 cm tall, what is Hussein's height?

8. A type of brass is an alloy of copper and zinc in the ratio 3 : 2. How much copper is alloyed with 18 kg of zinc?

9. Two friends buy a racehorse, and agree to contribute money in the ratio 3 : 5. If the smaller share is £15 000, what is the larger share?

10. A recipe for pastry tells the cook to use 6 oz of fat and 8 oz of flour. How much fat will be required for 12 oz of flour?

11. A motorist reckons that a journey of 100 miles costs him £9. Assuming that the cost per mile is constant, what will be the cost of a journey of 250 miles?

12. A car travels 6 miles on one litre of petrol. How far will it get on a tankful of 40 litres?

13. The weight of 5 cm^3 of a metal is 48 grams. What is the weight of 30 cm^3 of the metal?

14. The cost of 5 metres of material is £32. How much will 25 metres cost?

15. Look at the recipe for Greek-style mushrooms (Fig. 4.1). Write down the ingredients required if $1\frac{1}{2}$ lb of mushrooms are used.

```
DIMITRA'S GREEK
STYLE MUSHROOMS

½ lb mushrooms
3 tablespoons olive oil
1 small lemon
1 clove garlic
4 tablespoons parsley
```

Fig. 4.1

```
Mrs Brown's Scones

250 grams flour
2 teaspoons cream of tartar
1 teaspoon bicarbonate
50 grams butter
30 grams sugar
50 ml milk
```

Fig. 4.2

16. Mrs Brown's recipe (Fig. 4.2) gives the ingredients to make 10 large scones. Write down the quantities needed to make 25 large scones.

Division in a ratio

If an area of 27 acres is divided between two brothers in the ratio 2 : 7, the area is divided into 2 + 7, i.e. 9, equal portions. Each portion is 27 ÷ 9 = 3 acres. One brother gets two of these portions, i.e. 6 acres, and the other brother gets seven portions, i.e. 21 acres.

■ Example 4.9

A type of concrete is made from cement, sand and aggregate in the ratio 1 : 3 : 6. How much sand is needed for 40 kg of concrete?

Solution

This example shows that ratios can compare more than two quantities. The number of portions is 1 + 3 + 6, i.e. 10. Each portion is 40 kg ÷ 10 = 4 kg. Multiply 4 by 3 to give the quantity of sand.
12 kg of sand is needed.

Key point

- When a quantity is divided in the ratio 2 : 5, the first part is $\frac{2}{7}$ of the whole, not $\frac{2}{5}$.

EXERCISE 4F

1. In a school the ratio of boys to girls is 4 : 5. If there are 1800 pupils in total how many are boys?

2. An alloy consists of copper and zinc in the ratio 11 : 9. How much zinc is there in 30 kg of the alloy?

3. In a will money is left to Charles and Davina in the ratio 3 : 5. If the total sum of money is £24 000 how much does Davina get?

4. Mr Patel and Mr Shah start up a business. Their contributions are in the ratio 2 : 1.

   a) The cost of setting up the business is £600 000. How much does Mr Patel contribute?

   b) In the first year they make a profit of £30 000. If profits are divided in the same ratio, how much does Mr Shah get?

5. Susanna and Gail share a flat. As Susanna's room is larger, the rent is divided in the ratio 7 : 5. If the rent is £180 per week, how much does Gail pay?

6. A sum of money is left to three family members in the ratio 4 : 5 : 6. If the amount of money is £60 000, how much does each person get?

7. An exam is divided into Sections A, B and C. Marks are awarded in the ratio 5 : 3 : 2 for the three parts respectively. If the total marks are 200, how many are given for Section C?

8. The heights of the three storeys of a house are in the ratio 11 : 10 : 9. If the total height of the house is 9 m, how high is each storey?

Revision checklist

This chapter has revised:

4.1 Finding a percentage of a quantity. ❏

4.2 Expressing a quantity as a percentage of another, and finding a percentage change. ❏

4.3 Finding sums of money after a few years of compound interest. Using a reverse percentage to find the original value of a quantity, given its changed value. ❏

4.4 Finding the ratio between two quantities, and dividing a quantity in a ratio. ❏

CHAPTER 5 Powers

Chapter key points

- Don't confuse squaring and doubling:

    $5^2 \neq 2 \times 5$

    nor taking the square root and halving:

    $\sqrt{5} \neq \frac{1}{2} \times 5$

- Be careful when using indices. Don't confuse *taking powers* with *multiplying* or *dividing*. Watch out for:

    $7^3 \times 7^4 \neq 7^{12}$ $2^3 + 2^4 \neq 2^7$
    $3^3 \times 5^2 \neq 15^5$ $(2^5)^2 \neq 4^{10}$

- Remember that taking powers is done before other operations:

    $3 \times 2^2 \neq 6^2$

- In standard form there is only one digit to the left of the decimal point. 57×10^3 is not in standard form.
- Watch out for:

    $4 \times 10^6 \times 2 \times 10^5 \neq 8 \times 10^{30}$
    $2 \times 10^5 + 3 \times 10^5 \neq 5 \times 10^{10}$

- When using the EXP key of a calculator, don't key in 10. To enter 5×10^6, for example, the sequence is:

    | 5 | EXP | 6 |

    If you key in the following, your answer will be 10 times too large:

    | 5 | × | 1 | 0 | EXP | 6 |

Chapter 1 dealt with the four basic operations. Here we deal with a fifth operation, taking powers.

5.1 Squares, cubes, roots

The **square** of a number is the number multiplied by itself. It gives the area of a square. See Fig. 5.1.

$3 \times 3 = 3^2 = 9$

The **cube** of a number is the number multiplied by itself twice. It gives the volume of a cube. See Fig. 5.2.

$4 \times 4 \times 4 = 4^3 = 64$

The square of 7 is 49. The **square root** of 49 is 7.

$49 = 7^2 \quad 7 = \sqrt{49}$

A calculator can find square roots. Different calculators require different orders of keys. Probably one of the sequences below will work on your calculator. Find which one.

| 4 | 9 | √ | or | √ | 4 | 9 | = |

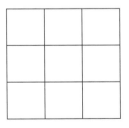

Fig. 5.1

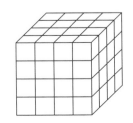

Fig. 5.2

POWERS

■ Example 5.1

Evaluate the following.

a) 7^2 b) 2^3 c) $\sqrt{100}$

Solution

a) Use the definition: $7 \times 7 = 49$.
   $7^2 = 49$
b) Use the definition: $2 \times 2 \times 2 = 8$.
   $2^3 = 8$
c) Note that $10^2 = 100$.
   $\sqrt{100} = 10$

■ Example 5.2

a) A chess board is a square divided into several smaller squares. There are eight squares along each side. How many smaller squares are there?
b) *Go* is a Japanese game played on a square board divided into smaller squares. There are 289 smaller squares. How many small squares are there along each side?

Solution

a) Square 8, obtaining 64.
   There are 64 smaller squares.
b) Find the square root of 289. This is 17.
   There are 17 smaller squares along each side.

Key point

■ Don't confuse squaring and doubling:

$$5^2 \neq 2 \times 5$$

nor taking the square root and halving:

$$\sqrt{5} \neq \tfrac{1}{2} \times 5$$

EXERCISE 5A

1 Evaluate the following.
   a) 3^2 b) 9^2 c) 5^2 d) 11^2

2 Evaluate the following.
   a) $\sqrt{25}$ b) $\sqrt{4}$ c) $\sqrt{64}$ d) $\sqrt{144}$

3 Evaluate the following.
   a) 3^3 b) 5^3 c) 4^3 d) 10^3

4 Evaluate the following.
   a) $(\tfrac{1}{3})^2$ b) $(\tfrac{1}{2})^3$ c) $\sqrt{\tfrac{1}{4}}$

5 A field is a square with side 50 m. What is its area?

6 A carpet is a square with side 4 feet. What is the area of the carpet?

7 The top of a table is a square of area 6400 cm²? What is the side of the top?

8 A square of cloth has area 169 square inches. What is the side of the square?

9 A cube has side 5 cm. What is the volume of the cube?

5.2 Indices

The *n*th **power** of a number consists of *n* copies of the number multiplied together. The number *n* is the **index**. The expression below is the fifth power of 2. The index is 5.

$$2 \times 2 \times 2 \times 2 \times 2 = 2^5 = 32$$

Zero and negative powers

Any positive number to the 0 power is 1.

$$4^0 = 1$$

If the index is negative, you can convert the negative power to 1 over the positive power:

$$4^{-2} = \frac{1}{4^2} = \tfrac{1}{16}$$

Multiplying and dividing powers

When powers are multiplied, the indices are added.

$$4^3 \times 4^5 = (4 \times 4 \times 4) \times (4 \times 4 \times 4 \times 4 \times 4) = 4^8$$

When powers are divided, the indices are subtracted.

$$5^7 \div 5^3 = \frac{5 \times 5 \times 5 \times 5 \times \cancel{5} \times \cancel{5} \times \cancel{5}}{\cancel{5} \times \cancel{5} \times \cancel{5}} = 5^4$$

Order of operations

Taking powers is done before multiplying or dividing.

$$5 \times 3^2 = 5 \times 9 = 45$$

To multiply first, use brackets.

$$(5 \times 3)^2 = 15^2 = 225$$

Powers with a calculator

You can use a scientific calculator to find powers. The sequence might be as follows.

| 5 | x^y | 3 | = | gives 5^3, i.e. **125**

| 5 | x^y | 3 | ± | = | gives 5^{-3}, i.e. **0.008**

■ Example 5.3

Evaluate the following.

a) 3^4 **b)** 5^{-2}

Solution

a) Use the definition: $3 \times 3 \times 3 \times 3 = 81$.
   $3^4 = 81$

b) Use the definition of a negative power. Take 1 over 5^2, i.e. 1 over 25.

$$5^{-2} = \frac{1}{5^2} = \tfrac{1}{25}$$

POWERS

Key points
- Be careful when using indices. Make sure you don't confuse *taking powers* with *multiplying* or *dividing*. Watch out for:

$$7^3 \times 7^4 \neq 7^{12} \qquad 2^3 + 2^4 \neq 2^7$$
$$3^3 \times 5^2 \neq 15^5 \qquad (2^5)^2 \neq 4^{10}$$

- Remember that taking powers is done before the other operations.

$$3 \times 2^2 \neq 6^2$$

EXERCISE 5B

1. Evaluate the following.
   a) 10^4
   b) 5^4
   c) 1^5
   d) 2^9

2. Evaluate the following.
   a) 2^{-2}
   b) 2^{-3}
   c) 1^{-3}
   d) 10^{-3}

3. Simplify the following.
   a) $2^3 \times 2^5$
   b) $3^2 \times 3^3$
   c) $5^8 \div 5^3$
   d) $2^5 \times 2^{-2}$
   e) $7^4 \div 7^{-3}$
   f) $8^5 \div 8^{-2}$

4. Use your calculator to find the following, giving your answer correct to 3 decimal places.
   a) $\sqrt{3}$
   b) $\sqrt{\frac{1}{2}}$
   c) $\sqrt{44}$
   d) $\sqrt{0.03}$

5. Find $\sqrt{10}$, giving your answer correct to:
   a) 1 decimal place
   b) 2 decimal places
   c) 5 decimal places

6. Evaluate the following, giving your answer correct to 2 decimal places.
   a) $3 + \sqrt{2}$
   b) $21 - \sqrt{3}$
   c) $\dfrac{\sqrt{5} + \sqrt{7}}{2}$
   d) $\dfrac{9}{\sqrt{7} - \sqrt{2}}$

7. A *kilobyte* is 2^{10}, i.e. 1024 bytes. A *gigabyte* is 2^{30} bytes. How many kilobytes are there in a gigabyte?

8. A *teravolt* is 10^{12} volts. A *microvolt* is 10^{-6} volts. How many microvolts are there in a teravolt?

9. Zeno has to evaluate the expression $\sqrt{(3.2 + 6.1)}$. He presses the sequence below on his calculator. Explain why he gets a wrong answer, and give a correct sequence.

   $\boxed{3}\,\boxed{.}\,\boxed{2}\,\boxed{+}\,\boxed{6}\,\boxed{.}\,\boxed{1}\,\boxed{\sqrt{}}\,\boxed{=}$

10. Gina wants to find the area of a circle with radius 3 cm, using the formula πr^2. She presses the following sequence of buttons. Explain what is wrong, and give a correct sequence.

    $\boxed{\pi}\,\boxed{\times}\,\boxed{3}\,\boxed{=}\,\boxed{x^2}$

5.3 Standard form

When a number is in **standard form**, there is only one digit to the left of the decimal point. The size of the number is given by a power of 10.

Standard form is used for very large and very small numbers. For example:

$$270\,000 = 2.7 \times 10^5 \qquad 0.000\,004\,3 = 4.3 \times 10^{-6}$$

You can use a scientific calculator for numbers in standard form. The sequences for the numbers above might be as follows.

| 2 | . | 7 | EXP | 5 | enters 2.7×10^5

| 4 | . | 3 | EXP | 6 | ± | enters 4.3×10^{-6}

Multiplication

When multiplying numbers in standard form, multiply the number parts and *add* the indices.

$$2 \times 10^7 \times 3 \times 10^8 = 6 \times 10^{15}$$

Division

When dividing numbers in standard form, divide the number parts and *subtract* the indices.

$$8.2 \times 10^9 \div 2 \times 10^4 = 4.1 \times 10^5$$

Addition and subtraction

When adding or subtracting numbers in standard form, first adjust the numbers so that they have the same power of 10. Then add the number parts.

$$5 \times 10^4 + 4 \times 10^5 = 0.5 \times 10^5 + 4 \times 10^5 = 4.5 \times 10^5$$

■ Example 5.4

Evaluate $7 \times 10^6 \times 3 \times 10^9$, leaving your answer in standard form.

Solution

Multiply 7 by 3, and add 6 and 9.
21×10^{15}
This is the correct number, but it is not in standard form. Change 21 to 2.1, and increase the power of 10 by 1.
$7 \times 10^6 \times 3 \times 10^9 = 2.1 \times 10^{16}$

■ Example 5.5

The star Sirius is 5×10^{13} km from Earth. How long will a spaceship travelling at 8000 km per second take to reach it?

Solution

Write 8000 in standard form as 8×10^3. Divide this into 5×10^{13}.
$5 \times 10^{13} \div 8 \times 10^3 = 0.625 \times 10^{10}$
Convert to standard form by changing 0.625 to 6.25 and subtracting 1 from the power of 10.
It will take 6.25×10^9 seconds.

■ Example 5.6

A ship weighs 6×10^7 kg when empty. What is its weight when it carries cargo of 3×10^6 kg?

Solution

Add these numbers. Adjust the second number so that it has the same power of 10 as the first number.
$3 \times 10^6 = 0.3 \times 10^7$
Now add.
$6 \times 10^7 + 3 \times 10^6 = 6 \times 10^7 + 0.3 \times 10^7 = 6.3 \times 10^7$
The weight is 6.3×10^7 kg.

© IT IS ILLEGAL TO PHOTOCOPY THIS PAGE

Key points

- In standard from, there is only one digit to the left of the decimal point. 57×10^3 is not in standard form.
- Watch out for:

$$4 \times 10^6 \times 2 \times 10^5 \neq 8 \times 10^{30}$$
$$2 \times 10^5 + 3 \times 10^5 \neq 5 \times 10^{10}$$

- When using the EXP key of a calculator, don't key in 10. To enter 5×10^6, for example, the sequence is:

  [5] [EXP] [6]

  If you key in the following, your answer will be 10 times too large:

  [5] [×] [1] [0] [EXP] [6]

EXERCISE 5C

1 Express the following in standard form.

a) 410 000 b) 901 000 000 c) 0.002 73 d) 0.000 007 2

2 Evaluate the following, leaving your answers in standard form:

a) $2 \times 10^6 \times 3 \times 10^4$ b) $3 \times 10^5 \times 4 \times 10^4$ c) $5 \times 10^8 \times 3 \times 10^7$

d) $3.6 \times 10^8 \div 2 \times 10^3$ e) $4.2 \times 10^5 \div 2 \times 10^8$ f) $1.2 \times 10^5 \div 4 \times 10^2$

g) $4 \times 10^{-6} \times 1.2 \times 10^{-8}$ h) $7 \times 10^{-12} \times 3 \times 10^{-5}$

i) $6.5 \times 10^6 \div 5 \times 10^{-10}$ j) $3 \times 10^8 + 2 \times 10^8$ k) $7 \times 10^6 + 2 \times 10^5$

l) $8 \times 10^7 - 5 \times 10^6$

3 A cubic kilometre of water has mass 1×10^{12} kg. Find the mass of water in a sea which has area 300 000 km² and depth 2 km.

4 The speed of light is 1.86×10^5 miles per second. How far does a beam of light travel in a year?
(This distance is a *light year*.)

5 The weight of a hydrogen atom is about 2×10^{-28} grams. How many atoms are there in 10 grams of hydrogen?

6 The population of a certain country is 1.2×10^8 people. The average annual income is 2×10^4 dollars. What is the total income of the country?

7 The computer magnate Gill Bates has total wealth three thousand million dollars. Write her wealth as a number in full and in standard form.

8 The population of the world is about 6×10^9 people. The area of the Isle of Wight is about 4×10^8 m².
If everyone went and stood on the Isle of Wight, how much room would each person get?

9 The populations of China and Japan are about 1.1×10^9 and 1.2×10^8 respectively. What is their combined population?

10 The Earth is about 1.5×10^8 km from the Sun. The planet Saturn is about 1.43×10^9 km from the Sun.
Suppose the Sun, the Earth and Saturn are in a straight line. What is the distance between Earth and Saturn:

a) when they are on directly opposite sides of the Sun

b) when they are on the same side of the Sun?

POWERS

Revision checklist

This chapter has revised:

5.1 The squares, cubes and square roots of numbers. ❏
5.2 Higher powers of numbers. Zero and negative powers. The arithmetic of powers, and the order of operations. Use of a calculator to find powers. ❏
5.3 Putting large or small numbers in standard form. The arithmetic of numbers in standard form. Use of a calculator for numbers in standard form. ❏

CHAPTER 6 Error and approximation

> **Chapter key points**
> - When you round a number, keep the zeros. 5132 to the nearest 100 is 5100, not 51.
> - Do the rounding at the end of a calculation. For example:
>
>   1.4 + 1.3 = 2.7 = 3 to the nearest whole number
>
>   If you do the rounding first, you may get the wrong answer:
>
>   1.4 + 1.3 = 1 + 1 = 2
>
> - The first significant figure is the first *non-zero* figure. So the first significant figure of 0.035 is 3 not 0.
> - After the first significant figure, zeros are significant. The first two significant figures of 508 are 5 and 0, not 5 and 8.
> - Suppose you are rounding 3.804 correct to 2 decimal places. After rounding, the digit in the second decimal place is 0. Leave it in to show the accuracy of the measurement – give the answer as 3.80, not 3.8.
> - Suppose a value is given as 2.3 correct to 1 decimal place. The upper limit on the true value is 2.35, not 2.4 or 2.34.
> - Checks of accuracy are very useful, but they don't *prove* that your answer is right – they may prove that it is wrong!
> - Don't give an answer that is more accurate than appropriate.

When you measure a length, a time, a weight, and so on, your result is not absolutely accurate. Your result is approximate. You should be aware of the possible error in your measurement, and of appropriate levels of accuracy in numerical calculations.

6.1 Rounding

We **round** a figure to a whole number by writing the nearest whole number. For example:

   73.9 is nearest to 74
   20.2 is nearest to 20

We round a figure to the nearest 100 by writing the nearest multiple of 100. For example:

   241 is nearest to 200
   683 is nearest to 700

■ Example 6.1

A temperature is measured as −5.7°C. Round this to the nearest whole number.

Solution
See Fig. 6.1. Notice that −5.7 is nearer −6 than −5.
Round to −6°C.

Fig. 6.1

Example 6.2

A census gives the population of a city as 413 650. Round this to:

a) the nearest 1000
b) the nearest 100.

Solution

a) 650 is nearer 1000 than 0.
   Round to 414 000.
b) 50 is halfway between 100 and 0. We could round it up or down. Usually we round up.
   Round to 413 700.

Example 6.3

A car is driven for 4 hours at 56 m.p.h. Correct to the nearest 10 miles, how far has it travelled?

Solution

Multiply the speed and the time.
$4 \times 56 = 224$ miles
This distance is closer to 220 than to 230.
To the nearest 10 miles, it has travelled 220 miles.

Key points

- When you round a number keep the zeros. 5132 to the nearest 100 is 5100, not 51.
- Do the rounding at the end of a calculation. For example:

    $1.4 + 1.3 = 2.7 = 3$ to the nearest whole number

    If you do the rounding first, you may get the wrong answer:

    $1.4 + 1.3 = 1 + 1 = 2$

EXERCISE 6A

1. Round the following numbers to the nearest whole number.

   a) 4.8 b) 5.2 c) 43.67 d) 65.5
   e) −8.25 f) −5.92 g) 0.83 h) 0.4
   i) $2\frac{1}{3}$ j) $\frac{7}{11}$

2. The weight of a ship is given as 23 847 114 kg. Round this to:

   a) the nearest 1000 kg
   b) the nearest 10 000 kg

3. Work out the following, giving your answer to the nearest whole number.

   a) 51.3 + 22.8 b) 1.5 + 7.2 + 1.9 c) 2.3 × 2.4
   d) 62.6 − 12.9 e) 62 ÷ 7 f) 21.6^2

4. What is the cost of 27 litres of petrol of £0.578 per litre? Give your answer correct to the nearest pound.

5. One kg of potatoes costs 37p. What is the cost of 86 kg of potatoes? Give your answer correct to the nearest pound.

6. It costs £2.90 to hire a tennis court for one hour. What is the cost per minute, given correct to the nearest penny?

7 A car is driven at 65 m.p.h. for $4\frac{1}{3}$ hours. How far has it gone? Give your answer correct to the nearest 10 miles.

8 I have to drive a distance of 300 miles. If my average speed is 47 m.p.h., how long will it take me? Give your answer correct to the nearest hour.

9 A lottery win of £100 000 is shared out equally between 7 people. How much does each get, given correct to the nearest £1000?

6.2 Decimal places and significant figures

The **first decimal place** of a number is the place after the decimal point. The second decimal place is the next place.

The **first significant figure** is the first non-zero digit. The second significant figure is the next digit, even if it is zero.

■ Example 6.4

The width of a building is found to be 5.38 m. Write this correct to 1 decimal place.

Solution
The digit in the first decimal place is 3. The next figure is 8, so round up.
The width is 5.4 m.

■ Example 6.5

A population is given as 34 648 000 people. Express this correct to 3 significant figures.

Solution
The third significant figure is 6. The figure to the right is 4, so round down.
The population is 34 600 000.

Key points
- The first significant figure is the first *non-zero* figure. So the first significant figure of 0.035 is 3, not 0.
- After the first significant figure, zeros are significant. The first two significant figures of 508 are 5 and 0, not 5 and 8.
- Suppose you are rounding 3.804 correct to two decimal places. After rounding, the digit in the second decimal place is 0. Leave it in, to show the accuracy of the measurement – give the answer as 3.80, not 3.8.

EXERCISE 6B

1 Round the following correct to 2 decimal places.

 a) 1.238 b) 4.123 c) 4.199 d) 0.002

2 Round 54.7592 correct to:

 a) 1 decimal place b) 2 decimal places c) 3 decimal places

3 Round the following correct to 3 significant figures.

 a) 2389 b) 309 231 c) 0.003 812

4   Round 1.2098 correct to:

   a) 3 significant figures   **b)** 2 significant figures   **c)** 1 significant figure

5   Evaluate the following, giving your answers correct to 2 decimal places.

   a) $2.613 + 6.339$ **b)** $7.4811 - 0.3733$ **c)** 7.915×3.622

6   The *density* of a substance is its mass divided by its volume. If 18 grams of a substance occupies 13 cm^3, what is its density? Give your answer correct to 3 decimal places.

7   Certain bolts weigh 1.9 grams each. What is the weight of 317 of these bolts? Give your answer correct to 2 significant figures.

8   Fig. 6.2 shows the display of a calculator after 32 has been divided by 13. Write this number correct to:

   a) 3 significant figures   **b)** 4 decimal places $\boxed{2.461538462}$

   Fig. 6.2

Greatest and least values

If a number is given as 2.7 correct to 1 decimal place, the maximum error is 0.05, i.e. 5 in the second decimal place. In Fig. 6.3, the number could lie anywhere between 2.65 and 2.75.

Fig. 6.3

If a number is given as 23 000 correct to 2 significant figures, the maximum error is 500.

■ Example 6.6

A length is given as 2.34 m correct to 2 decimal places. Between what values does the length lie?

Solution

The greatest error is 5 in the third decimal place, i.e. 0.005. Add and subtract this from 2.34.
The length lies between 2.335 m and 2.345 m.

Key point

■ Suppose a value is given as 2.3 correct to 1 decimal place. The upper limit on the true value is 2.35, not 2.4 or 2.34.

EXERCISE 6C

1   A person's weight is 64 kg, correct to the nearest whole number. Between what values does the weight lie?

2   The population of a certain country is given as 48 000 000, correct to the nearest million. Between what values does the population lie?

3 A temperature is measured as 17°C, given correct to the nearest whole number. What is the greatest possible value of the temperature?

4 A length is measured as 37.4 cm, correct to 1 decimal place. Between what limits does the length lie?

5 The area of a farm is 47.3 hectares, correct to 3 significant figures. What is the least possible area of the farm?

6 The speed of a car is 27.32 m/s, correct to 2 decimal places. What are the greatest possible speed and the least possible speed?

6.3 Checking by approximation

It is easy to make a mistake when calculating. Check by rounding each figure correct to 1 significant figure and using mental arithmetic.

■ Example 6.7

Evaluate 3.295×1.858. Check your answer by approximation.

Solution
By calculator, obtain 6.12211.
To check, round both terms correct to 1 significant figure, then multiply.
$3 \times 2 = 6$
6 is close to the value found by calculator. This checks the answer.
$3.295 \times 1.858 = 6.12211$

Key point

■ Checks of accuracy are very useful, but they don't *prove* that your answer is right – they may prove that it is wrong!

EXERCISE 6D

1 Evaluate the following. Make checks to verify your answers.

   a) $5.1 + 7.8$ b) $2.8 + 3.1$ c) $6.17 - 1.31$
   d) $66.29 - 41.71$ e) 7.2×4.8 f) 12.2×3.9
   g) $6.3 \div 2.8$ h) $26.7 \div 2.6$ i) $(1.44 \times 5.92) \div 1.77$
   j) $(7.25 + 6.11) \times (3.72 + 5.29)$

2 A tank of weight 163 kg contains 743 kg of water. What is the approximate total weight?

3 In France, you buy two items, one for 38F and one for 73F. Approximately, how much change do you expect from a 200F note?

4 Twenty-three crates each weigh 178 kg. What is the approximate total weight?

5 I buy 29 litres of petrol at 57.3p per litre. About how much will I pay?

6 Three people go to Spain, taking £325, £216 and £376 respectively. They pool their money and change it to pesetas at 212 pesetas per £. Roughly how much should they get?

7 An electricity bill is made up of a standing charge of £13.36, 1313 units at the low rate of 2.7p per unit, and 1709 units at the high rate of 7.4p per unit. What is the approximate size of the total bill?

6.4 Appropriate levels of accuracy

An answer should be given to an appropriate level of accuracy. Don't give an answer with too many decimal places or too many significant figures.

■ Example 6.8

A city of about 500 000 people is divided into seven wards of approximately equal size. How many people are there in each ward?

Solution
Divide 500 000 by 7, obtaining 71 429 people. This is too accurate – you started with a number that was written to 1 significant figure. Round to the same degree of accuracy, i.e. 1 significant figure.
Each ward contains about 70 000 people.

Key point
■ Don't give an answer that is more accurate than appropriate.

EXERCISE 6E

1. Over 28 days, there was rainfall of approximately 3 inches. What was the average rainfall per day? Give your answer to an appropriate degree of accuracy.

2. Certain tiles cost 37p each. How much will it cost for about 450 tiles?

3. The price of a new Siesta car is about £12 000, but a discount of 12% is given. What is the cost after the discount?

4. The radius of the Earth is 6 400 000 metres, correct to 2 significant figures. Find the circumference of the Earth, giving your answer to an appropriate degree of accuracy. (The circumference of a circle of radius r is $2\pi r$).

5. What would be appropriate levels of accuracy for the following measurements?
   a) The population of the UK.
   b) The world record for the 100 m sprint.
   c) Your fastest time for the 100 m sprint.
   d) The exchange rate in August of the $ to the £.
   e) The price of your family's house or flat.
   f) The top speed of a new Siesta car.

Revision checklist

This chapter has revised:

6.1 Rounding numbers to the nearest whole number, the nearest 10, etc. ❏
6.2 Decimal places and significant figures. The limits of a measurement given to a certain number of decimal places or significant figures. ❏
6.3 Using an approximation to check a calculation. ❏
6.4 Giving the result of a measurement or calculation to an appropriate level of accuracy. ❏

CHAPTER 7 *Practical arithmetic*

Chapter key points
- If some items are 40p each and others are 60p each, then the average cost is not necessarily 50p. To find the average cost, divide the total cost by the total number of items.
- Clock times are not decimals. There are 60 minutes in an hour, not 100. So 0840 is 20 minutes to 9, not 60 minutes to 9.
- The 24 hour clock adds on 12 hours to afternoon times. Hence 1500 represents 3p.m., not 5p.m.

In this chapter we deal with practical problems involving money, averages, and so on.

7.1 Rates and averages

A **rate** is one quantity divided by another. For example cost per kilogram, wages per hour, miles per gallon.

If you buy several items, the **average** cost per item is the total cost divided by the number of items.

■ Example 7.1

I buy petrol at £0.58 per litre.

a) How much will 12 litres cost?
b) How much petrol can I buy for £10?

Solution

a) Multiply 0.58 by 12, obtaining 6.96.
   12 litres will cost £6.96.
b) Divide 10 by 0.58, obtaining 17.24.
   £10 will buy 17.24 litres.

■ Example 7.2

I can buy white paint either in 2 litre cans for £8.60 each, or in 5 litre cans for £21.20 each (Fig. 7.1). In which can is the paint cheaper?

Solution
The price of 'Brilliant White' per litre is £8.60 ÷ 2, i.e. £4.30. The price of 'Shining White' per litre is £21.20 ÷ 5, i.e. £4.24.
'Shining White' paint in the 5 litre cans is cheaper.

Fig. 7.1

■ Example 7.3

I buy 12 shirts at £15 each, and 8 shirts at £18 each. What is the average cost of the shirts?

Solution
The total cost is £(12 × 15 + 8 × 18), i.e. £324. The total number of shirts is 20. Divide 324 by 20, obtaining 16.2.
The average cost is £16.20 per shirt.

PRACTICAL ARITHMETIC

Key point

- If some items are 40p each and others are 60p each, then the average cost is not necessarily 50p. To find the average cost, divide the total cost by the total number of items.

EXERCISE 7A

1. **a)** A 6 lb joint of beef costs £15. What is the cost per £?
   b) I use 20 litres of petrol in covering 180 km. What is the fuel economy, in km per litre?
   c) A man earns £420 for a 35 hour week. What are his wages per hour?

2. What is the cost of the following?
   a) 15 kg of cheese at £6 per kg.
   b) An eight-minute phone call at 0.35p per minute.

3. **a)** With £24, how much petrol could I buy at 60p per litre?
   b) With £2.40, what quantity of beans could I buy at £1.50 per pound?

4. A bucket has a hole in it. Initially the bucket contains 5000 cm^3 of water, and water drips out at a steady rate of 20 cm^3 per minute.
   a) When will the bucket be empty?
   b) After two hours, how much water will be left in the bucket?

5. A school buys 150 copies of a book at £5.50 each, and 100 copies of another book at £6.00 each. What is the average cost of the books?

6. The fuel consumption of a car is 15 km per litre of petrol. How much fuel does it take to cover 330 km?

7. Over a journey of 320 km, a car uses 25 litres of petrol. What is the fuel economy of the car in km per litre?

8. An electricity bill consists of a standing charge of £16 and a rate per unit of 3.8p. If a household uses 2560 units, what is the total bill?

9. An overseas telephone call lasts 3 minutes 40 seconds. How much does it cost, at a rate of 1.05p per second?

10. The annual subscription for a health club, running from 1st January to 31st December, is £450. Jenny joins on 15th May. If her annual subscription is reduced proportionally, how much does she pay?

11. Fig. 7.2 shows two bottles of white spirit. Which offers better value?

12. The advertisements in Fig. 7.3 show two ways of buying a car. Which way is cheaper?

Fig. 7.2

Economy!	*Supersaver!*
£2000 deposit, then £200 per week for 50 weeks!	No deposit, then £400 per week for 29 weeks!

Fig. 7.3

7.2 Money matters

- The money we earn is paid either as a **wage** or as a **salary**. A wage is paid per hour or per week. A salary is paid per month or per year.
- **Overtime** is extra hours of work. Overtime is often paid at 'time and a half', which is $1\frac{1}{2}$ times the usual hourly rate.
- **Income tax** is a tax taking a percentage of income.
- **Value added tax** (**VAT**) is a tax taking a percentage of the price of goods.
- When goods are bought on **hire-purchase**, there is usually a down payment followed by several monthly or weekly payments.

■ Example 7.4

George is paid at an hourly rate of £6.50, for a week of 38 hours. Overtime is paid at 'time and a half'. How much does he receive for a week in which he worked 44 hours?

Solution

He worked 6 hours overtime. Each overtime hour is paid at £6.50 × $1\frac{1}{2}$, i.e. £9.75.
The total amount he is paid is 38 × £6.50 + 6 × £9.75 = £305.50.
He receives £305.50.

EXERCISE 7B

1. What are the yearly incomes corresponding to the following?

   a) £1800 per month   b) £225 per week
   c) £8.20 per hour for a 40 hour week

2. What are the weekly incomes corresponding to the following?

   a) £5.80 per hour for 38 hours   b) £16 380 per year

3. Roger is paid £244.80 for a 36 hour week. What is the hourly rate?

4. Tom earns £8.70 per hour for a basic 36 hour week. Overtime is paid at 'time and a half'. How much does he receive for a week in which he worked 45 hours?

5. Babs is paid £6.30 per hour for a 30 hour week. Overtime is at 'time and a half'. If she is paid £226.80 for a week's work, how much overtime has she done?

6. Jane sees two jobs advertised. One is for a keyboard operator at £12 300 per year, the other for a shop worker at £240 per week. Which job is better paid?

7. A simple income tax scheme is as follows. Each citizen has a tax-free allowance. The tax is 25% of the citizen's income above the allowance.

   a) A single woman has an allowance of £3000 and an income of £15 500. How much tax does she pay?
   b) A married man with an allowance of £5500 pays £4400 tax. What is his income before tax?

8. At the time of writing the rate of VAT is $17\frac{1}{2}$%. Find the VAT on the following:

   a) a computer costing £800   b) car repair work costing £320

9. A car is bought on hire purchase, for £1850 down payment and 36 monthly payments of £120. What is the total cost?

10. A three-piece suite can be bought either for £860 cash, or on hire purchase for a down payment of £200 and 12 monthly payments of £60. How much more does it cost if it is bought by hire purchase?

7.3 Tables and charts

Tables and charts give information about train times, distances between towns, prices of hotels, and so on.

The times of aeroplanes or trains are often given in the **24 hour clock** system. For an afternoon time, add on 12 hours to the time given by a 12 hour clock. For example:

6p.m. (12 hour clock) = 1800 (24 hour clock)

Example 7.5

Fig. 7.4 shows part of a timetable for trains from London Charing Cross to Ore.

a) If I take the 0940 at Charing Cross, how long will it take to get to Battle?
b) I want to go from Wadhurst to Hastings, arriving before midday. When should I be at Wadhurst station?

Monday to Friday

		4	22	2	22	23	4	22	2	22	4	26
						B						C
London Charing Cross	dep	0855	0910	0930	0940	....	1000	1010	1030	1040	1100	....
London Waterloo East	dep	0858	0913	0933	0943	....	1003	1013	1033	1043	1103	....
London Cannon Street	dep	*0843*	*0902*	*0923*	*0938*	0942	*0953*	*1008*	*1023*	*1038*	*1053*	....
London Bridge	dep	....	0918	0937	0947	0947	1007	1017	1037	1047	1107	....
Orpington	dep	0923	0937	*0937*	1003	1009	1023	1033	....	1103	1123	....
Chelsfield	dep	....	....	*0932*	....	....	....	*1028*	....	....	....	....
Knockholt	dep	....	....	*0934*	....	....	....	*1031*	....	....	....	....
Dunton Green	dep	....	....	*0939*	....	....	....	*1036*	....	....	....	....
Sevenoaks	dep	0932	0946	1001	1013	1018	1033	1043	1101	1113	1133	....
Hildenborough	dep	....	....	....	1019	....	....	....	....	1119	....	....
Tonbridge	arr	0941	0955	1009	1023	1027	1041	1051	1109	1123	1141	....
Tonbridge	dep	....	0955	....	1023	1038	....	1054	....	1123	....	1145
High Brooms	dep	....	1001	....	1029	1044	....	1100	....	1129	....	1151
Tunbridge Wells	arr	....	1005	....	1033	1047	....	1103	....	1133	....	1154
Groombridge	arr	....	....	....	*1057*	....	....	....	....	*1157*	....	....
Tunbridge Wells	dep	....	1005	....	1033	1048	....	1104	....	1133	....	....
Frant	dep	....	....	....	1037	....	....	....	....	1137	....	....
Wadhurst	dep	....	1012	....	1041	....	....	1111	....	1141	....	....
Stonegate	dep	....	....	....	1047	....	....	....	....	1147	....	....
Etchingham	dep	....	....	....	1052	....	....	....	....	1152	....	....
Robertsbridge	dep	....	....	....	1056	....	....	....	....	1156	....	....
Battle	dep	....	1029	....	1104	1111	....	1127	....	1204	....	....
Crowhurst	dep	....	....	....	1107	....	....	....	....	1207	....	....
West St Leonards	dep	....	....	....	1112	1120	....	....	....	1212	....	....
St Leonards Warrior Square	dep	....	1039	....	1115	1123	....	1137	....	1215	....	....
Hastings	arr	....	1041	....	1118	1126	....	1140	....	1218	....	....
Ore	arr	....	....	....	....	*1137*	....	....	....	*1237*	....	....

Fig. 7.4 With permission from Connex South Eastern.

Solution

a) The time taken is the difference between the time of departure from Charing Cross and the time of arrival at Battle. The train leaves at 40 minutes past 9 and arrives at 4 minutes past 11.
It takes 20 minutes from 0940 to 1000.
It takes 60 minutes from 1000 to 1100.
It takes 4 minutes from 1100 to 1104.
Time taken = 84 minutes.

b) I must take the train which arrives in Hastings at 1140. This leaves Wadhurst at 1111.
I must be at Wadhurst before 11 minutes past 11.

Key points

- Clock times are not decimals. There are 60 minutes in an hour, not 100. So 0840 is 20 minutes to 9, not 60 minutes to 9.
- The 24 hour clock adds on 12 hours to afternoon times. Hence 1500 represents 3p.m., not 5p.m.

EXERCISE 7C

1 Convert the following times to 24 hour clock times.

a) 6a.m. **b)** 5p.m. **c)** 8.30p.m.

2 Convert the following times to 12 hour clock times.

a) 0800 **b)** 1400 **c)** 2330

3 A cross-channel ferry leaves Caen in France at 2300 and arrives at Portsmouth at 0545. How long has the crossing taken?

4 The calendar for 1997 is shown in Fig. 7.5. At a college, each term begins on a Monday and ends on a Friday.

a) The first term of the year began on 6th January and lasted for 11 weeks. When did it end?
b) The last term began on 15th September and ended on 12th December. How long was the term?

1997

January						
Mon.	.	6	13	20	27	.
Tues.	.	7	14	21	28	.
Wed.	[1]	8	15	22	29	.
Thur.	2	9	16	23	30	.
Fri.	3	10	17	24	31	.
Sat.	4	11	18	25	.	.
Sun.	5	12	19	26	.	.

February						
Mon.	.	3	10	17	24	.
Tues.	.	4	11	18	25	.
Wed.	.	5	12	19	26	.
Thur.	.	6	13	20	27	.
Fri.	.	7	14	21	28	.
Sat.	1	8	15	22	.	.
Sun.	2	9	16	23	.	.

March						
Mon.	.	3	10	17	24	[31]
Tues.	.	4	11	18	25	.
Wed.	.	5	12	19	26	.
Thur.	.	6	13	20	27	.
Fri.	.	7	14	21	[28]	.
Sat.	1	8	15	22	29	.
Sun.	2	9	16	23	30	.

April						
Mon.	.	7	14	21	28	.
Tues.	1	8	15	22	29	.
Wed.	2	9	16	23	30	.
Thur.	3	10	17	24	.	.
Fri.	4	11	18	25	.	.
Sat.	5	12	19	26	.	.
Sun.	6	13	20	27	.	.

May						
Mon.	.	[5]	12	19	[26]	.
Tues.	.	6	13	20	27	.
Wed.	.	7	14	21	28	.
Thur.	1	8	15	22	29	.
Fri.	2	9	16	23	30	.
Sat.	3	10	17	24	31	.
Sun.	4	11	18	25	.	.

June						
Mon.	.	2	9	16	23	30
Tues.	.	3	10	17	24	.
Wed.	.	4	11	18	25	.
Thur.	.	5	12	19	26	.
Fri.	.	6	13	20	27	.
Sat.	.	7	14	21	28	.
Sun.	1	8	15	22	29	.

July						
Mon.	.	7	14	21	28	.
Tues.	1	8	15	22	29	.
Wed.	2	9	16	23	30	.
Thur.	3	10	17	24	31	.
Fri.	4	11	18	25	.	.
Sat.	5	12	19	26	.	.
Sun.	6	13	20	27	.	.

August						
Mon.	.	4	11	18	[25]	.
Tues.	.	5	12	19	26	.
Wed.	.	6	13	20	27	.
Thur.	.	7	14	21	28	.
Fri.	1	8	15	22	29	.
Sat.	2	9	16	23	30	.
Sun.	3	10	17	24	31	.

September						
Mon.	1	8	15	22	29	.
Tues.	2	9	16	23	30	.
Wed.	3	10	17	24	.	.
Thur.	4	11	18	25	.	.
Fri.	5	12	19	26	.	.
Sat.	6	13	20	27	.	.
Sun.	7	14	21	28	.	.

October						
Mon.	.	6	13	20	27	.
Tues.	.	7	14	21	28	.
Wed.	1	8	15	22	29	.
Thur.	2	9	16	23	30	.
Fri.	3	10	17	24	31	.
Sat.	4	11	18	25	.	.
Sun.	5	12	19	26	.	.

November						
Mon.	.	3	10	17	24	.
Tues.	.	4	11	18	25	.
Wed.	.	5	12	19	26	.
Thur.	.	6	13	20	27	.
Fri.	.	7	14	21	28	.
Sat.	1	8	15	22	29	.
Sun.	2	9	16	23	30	.

December						
Mon.	1	8	15	22	29	.
Tues.	2	9	16	23	30	.
Wed.	3	10	17	24	31	.
Thur.	4	11	18	[25]	.	.
Fri.	5	12	19	[26]	.	.
Sat.	6	13	20	27	.	.
Sun.	7	14	21	28	.	.

Fig. 7.5

5 The chart in Fig. 7.6 shows the distances, in km, between several towns in France.

	Blois	Chartres	Le Mans	Orleans	Tours
Chartres	92				
Le Mans	109	107			
Orleans	49	73	138		
Tours	58	138	80	107	

Fig. 7.6

Use it to find:

a) the distance between Chartres and Le Mans
b) the distance from Tours to Chartres if one goes via Orleans.

6 Refer to the train timetable in Example 7.5 (Fig. 7.4).

a) If I catch the 0913 at Waterloo East, how long will it take to reach St Leonards Warrior Square?
b) I catch the 1113 at Sevenoaks. Where am I after 54 minutes?

7 The tables in Fig. 7.7 show the prices of two hotels in Italy. The Robinson family consists of Mr and Mrs Robinson, Darren (10), Christine (7) and Amy (4). They are going for a two-week holiday. How much would it cost them to stay in each of the hotels?

Hotel Rivoli			
Cost per person	*per week*	*per 14 days*	*Reductions per day*
	£350	£620	2–5 years £2 6–9 years £1.20

Hotel Pompeii			
Cost per person	*per week*	*per 14 days*	*Reductions per week*
	£440	£770	2–5 years £20 6–16 years £16

Fig. 7.7

8 Fig. 7.8 is a listing of the television programmes on a certain channel early one evening.

a) How long does the film last?
b) Gregory starts watching television at 6.15, and switches off $1\frac{3}{4}$ hours later. When does he switch off? How much of the film did he miss?

5.55 News
6.15 Weather
6.20 The Barney Show
7.05 Film: *The House with Green Shutters*
8.35 Regional news roundup

Fig. 7.8

Revision checklist

This chapter has revised:

7.1 Rates and averages, obtained by dividing one quantity by another. ❑
7.2 Money matters concerning pay, tax and hire purchase. ❑
7.3 Tables and charts for train times, distances, prices, and so on. ❑

Mixed exercise 1

1. When a space rocket is fired, the moment of launch is timed at 0 seconds.

   a) What is the time 4 seconds before launch?
   b) What is the time 20 seconds after −5 seconds?

2. A computer file of length 2 700 000 bytes is downloaded through the telephone line. If the rate of transmission is 600 bytes per second, how long will it take? Give your answer in minutes.

3. Jason downloads a photograph from the Internet. He starts at 1105, and after 6 minutes the unloading is 30% complete. When will it be finished?

4. The capacity of a floppy disk is 1400 kB. How much room is left if it contains files of total length 382 kB?

5. A professional tennis player normally gets 55% of her first serves in. In a match in which she served for 80 points, how many first serves would you expect that she got in?

6. When measuring angles, a minute is $\frac{1}{60}$ of a degree. A second is $\frac{1}{60}$ of a minute. What fraction of a degree is a second?

7. a) An alloy contains copper and zinc in the ratio 7 : 2. How much copper is alloyed with 40 kg of zinc?
   b) An ingot of another alloy contains 12 kg of copper and 8 kg of tin. What is the ratio of copper to tin? Give your answer in its simplest form.

8. a) Write the number five thousand and forty-eight in figures.
   b) Write 7033 in words.

9. Evaluate:

   a) $3 \times 4 + 7$ b) $3 \times (4 + 7)$

10. Jim has five £10 notes and seven £5 notes. He wants to change them into French Francs at 9FF per £.

    a) How much should he get?
    b) He uses the following sequence on a calculator. Explain why it is wrong.

    $\boxed{5} \boxed{\times} \boxed{1} \boxed{0} \boxed{+} \boxed{7} \boxed{\times} \boxed{5} \boxed{\times} \boxed{9} \boxed{=}$

    c) Write out a correct sequence of calculator keys.

11. The population of a certain country is 40 000 000. A special television programme was watched by 15 000 000 people. What percentage of the population watched the programme?

12. A journey of 200 miles takes 7 hours. What was the average speed of the journey? Give your answer correct to the nearest 10 m.p.h.

13. A candle is 30 cm high. It burns at a steady rate, so that it is finished after $1\frac{1}{2}$ hours.

    a) What is the rate of burning, in cm per minute?
    b) After how long will it be 10 cm high?
    c) What is its height after 12 minutes?

14. Arrange the following in increasing order.

    $\frac{1}{4}$ 20%   0.22 $\frac{2}{7}$

15. Fig. M1.1 shows two cans of oil: one costing £12.60 for 10 litres, the other costing £7.10 for 6 litres. Which is better value?

16. You have £385 in a building society account. You draw out £127.

    a) How much is left in the account?
    b) How could you perform a rough check that your answer is correct?

17. The area of a square sheet of board is 25 square feet. What is the side of the square?

Fig. M1.1

18 A woman earning £24 600 is given a 9% salary increase. How much does she earn after the increase?

19 For a weekend coach tour, 28 people each pay £46. Without using a calculator, evaluate the total amount they pay. Show all your working.

20 Fig. M1.2 shows the ingredients for avocado soup. Write down the ingredients if three avocados are used.

21 The manufacturers of Thinkwik slimming food claim that their diet will lead to a weight loss of 2 lb per week. Assume that this claim is correct.

  a) How much weight will be lost in 6 weeks?
  b) How long will it take to lose 1 stone? (1 stone = 14 lb)

Charles follows this diet, starting on 1st January. On that day he weighs 12 stones. You may find it helpful to use the calendar on page 44.

  c) What will he weigh on 12th February?
  d) When will he weigh 10 stone 12 lb?

2 ripe avocados
1 pint chicken stock
$\frac{1}{4}$ pint cream
juice of $\frac{1}{2}$ lemon

Fig. M1.2

22 In a class of 25 children, 15 are boys. Express the proportion of boys as a fraction in its simplest terms.

23 You are given the numbers 2, 8, 7, 17, 18. From these numbers write down:

  a) a cube number   **b)** a multiple of 6   **c)** a factor of 21

24 In a china factory, a quarter of the produce is flawed in the manufacture, and a tenth of the produce is broken before it reaches the shops. What fraction of the produce is sold in the shops?

The following questions are suitable for Intermediate level.

25 Without using a calculator or writing down any steps, evaluate $200\,000 \times 40\,000$.

26 A new town was founded in 1990 with a population of 100 000. Every year the population has grown by 11%. What was the population after 3 years?

27 The voltage of an electricity supply is 230 volts, to the nearest 10 volts. What are the limits between which the supply must lie?

28 The final match of a World Cup competition was watched on television by 450 000 000 people.

  a) Write this number in standard form.
  b) The final was heard on radio by 7×10^7 people. How many people either watched or heard the match? Give your answer in standard form.

29 Evaluate the following.

  a) $(-5) \times (-2) \times (-3)$ **b)** $(3 - 7) \div (17 - 19)$

30 A chemical compound contains hydrogen and carbon in the ratio 3 : 1.

  a) How much carbon is there in 24 kg of the compound?
  b) How much hydrogen is there in 1.6×10^{-8} kg of the compound? Give your answer in standard form.

Section 2
ALGEBRA

CHAPTER 8 Algebraic expressions

Chapter key points

- Don't ignore brackets in algebraic expressions:
  $$2(x + y) \neq 2x + y$$
- Be careful with negative numbers when substituting in algebraic expressions. Remember the rules:

  minus times minus is plus $- \times - = +$
  plus times plus is plus $+ \times + = +$
  minus times plus is minus $- \times + = -$

- The sequence 2, 7, 12, 17, 22, ... goes up in steps of 5. The nth term involves $5n$, not $n + 5$. The nth term is $5n - 3$.

An **algebraic expression** contains letters as well as numbers. An example is:

$$1.8C + 32$$

This formula converts temperature in Celsius to Fahrenheit. It is true for every possible temperature. The letter C can stand for any possible number.

8.1 Substitution

In an algebraic expression, **substitution** of numbers for letters gives the numerical value of the expression.

■ Example 8.1
Find the value of $1.8C + 32$ when $C = 25$.

Solution
Put $C = 25$ in the expression. The result is $1.8 \times 25 + 32$, i.e. 77.
The value is 77.

■ Example 8.2
The formula

$$\frac{1}{f} = \frac{1}{v} - \frac{1}{u}$$

is used for lenses. Find f when $v = 4$ and $u = 5$.

Solution
Put $v = 4$ and $u = 5$.

$$\frac{1}{f} = \frac{1}{4} - \frac{1}{5} = \frac{1}{20}$$

As $\frac{1}{f}$ is equal to $\frac{1}{20}$, f itself is equal to 20.

■ Example 8.3
The formula $c = \sqrt{a^2 + b^2}$ gives the longest side of a right-angled triangle. Find c when $a = 6$ and $b = 8$.

Solution
Put $a = 6$ and $b = 8$.
$c = \sqrt{6^2 + 8^2} = \sqrt{36 + 64} = \sqrt{100} = 10$

Key points

- Don't ignore brackets:

    $2(x + y) \neq 2x + y$

- Be careful with negative numbers. Remember the rules:

    minus times minus is plus $\quad - \times - = +$
    plus times plus is plus $\quad\;\, + \times + = +$
    minus times plus is minus $\quad - \times + = -$

EXERCISE 8A

1. If $y = 4x + 1$, find y when $x = 3$.

2. Ohm's Law connects current I, voltage V and resistance R by the formula $I = \dfrac{V}{R}$. Find I when $V = 12$ and $R = 8$.

3. The formula $v = u + at$ gives velocity v in terms of acceleration a, time t and initial velocity u. Find v when $a = 10$, $t = 3$, $u = 5$.

4. The area of a rectangle is given by $A = lb$. Find A when $l = 4$ and $b = 5$.

5. The area of a triangle is given by $A = \frac{1}{2}bh$. Find A when $b = 6$ and $h = 9$.

6. The area of a circle is given by $A = \pi r^2$. Find A when $r = 14$, taking π to be $\frac{22}{7}$.

7. The formula $I = PRT/100$ gives the interest I when a sum of P is invested at a percentage rate R for a time T. Find I when $P = 2000$, $R = 5$ and $T = 3$.

8. The formula $A = \frac{1}{2}h(x + y)$ gives the area of a trapezium. Find A when $h = 12$, $x = 5$ and $y = 4$.

9. Evaluate $5x + 7y$ when $x = 3$ and $y = 2$.

10. The formula $s = ut + \frac{1}{2}at^2$ gives the distance s after time t, under acceleration a and initial velocity u. Find s when $u = 7$, $t = 3$ and $a = 10$.

11. The formula $PVT = k$ relates gas pressure P, volume V and temperature T with a constant k. Find k when $P = 30$, $V = 45$, $T = 300$.

12. If the first n numbers are added, their sum is $\frac{1}{2}n(n + 1)$. Find the sum of the first 19 numbers.

13. The volume of a cone is given by $V = \frac{1}{3}\pi r^2 h$. Taking π to be $\frac{22}{7}$, find the volume of a cone for which $r = 3$ and $h = 14$.

14. Use the formula

    $$z = \frac{2 + w}{1 - w}$$

    to find z when $w = \frac{2}{3}$.

15. Use the formula

    $$\frac{1}{r} = \frac{1}{u} - \frac{1}{v}$$

    to find r when $u = 2$ and $v = 3$.

16. Two sides of a triangle, a and b, are connected by $a = \sqrt{625 - b^2}$. Find a when $b = 24$.

50 ALGEBRAIC EXPRESSIONS

8.2 Making formulae

When making a formula, apply operations to letters in the same way as you apply them to numbers.

■ Example 8.4

Anita weighs x kg. Bert is 5 kg heavier than Anita.
a) What is Bert's weight?
b) What is their total weight?

Solution
a) Add 5 to Anita's weight.
   Bert weighs $x + 5$ kg.
b) Add x and $x + 5$.
   Their total weight is $2x + 5$ kg.

■ Example 8.5

Cheap tickets for a concert cost £x and expensive tickets cost £y. What is the total cost of five cheap tickets and three expensive tickets?

Solution
The cheap tickets cost $5 \times £x$ and the expensive tickets cost $3 \times £y$. Add these.
The total cost is £$(5x + 3y)$.

EXERCISE 8B

1. James has £x. Kirsty has £20 more than James. How much does Kirsty have? How much do they have together?

2. A film lasts for x minutes. When it is shown on television, commercials lasting y minutes are shown during it. How long is the total programme?

3. The temperature at midday was t°C. At midnight, the temperature had fallen by 10°C. What was the temperature at midnight?

4. The angles of a triangle add up to 180°. Two of the angles are x° and y°. What is the third angle?

5. What is the cost of 10 books at £x each?

6. What is the cost of y books at £6 each?

7. A rectangular field is 100 metres long and x metres broad. What is the area of the field?

8. A win of £W is shared equally between 10 people. How much is each share?

9. A journey of m miles takes 3 hours. What is the average speed of the journey?

10. The value of x is 10 greater than the value of y. Write down x in terms of y.

11. A tennis match starts at 2 o'clock and ends at x minutes past 3 o'clock. How long has the match lasted?

12. The cost of hiring a van is a down-payment of £40 together with £10 per hour. How much does it cost to hire the van for x hours?

13. Bill asks Bryony: think of a number, multiply by 3, then add 4. Bryony's original number was n. What did she end up with?

14. Cakes cost x p each and buns cost y p each. What is the total cost of five cakes and seven buns?

15. A shop orders x pairs of shoes at £30 each and y pairs of boots at £40 each. What is the total cost?

8.3 Sequences

A **number sequence** is a list of numbers following a pattern. Some examples are:

2, 5, 8, 11, 14, ...   (adding 3 to the previous term)
2, 6, 18, 54, 162, ...   (multiplying the previous term by 3)
1, 2, 4, 7, 11, ...   (the differences between terms increasing by 1)

A **shape sequence** is a set of shapes following a pattern. It is often linked with a number sequence. Fig. 8.1 shows a shape sequence.

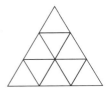

Fig. 8.1

■ Example 8.6
Continue the sequence 1, 2, 4, 7, 11 for two more terms.

Solution
The differences between the terms are 1, 2, 3, 4. Add 5 to 11 for the next term, 16. Add 6 to 16 for the term after that.
The next two terms are 16 and 22.

■ Example 8.7
The sequence 1, 4, 10, 22 is such that each term is found by adding one to the previous term and doubling.
a) Find the next term.
b) What is the term after x?

Solution
a) Add 1 to 22, then double.
   The next term is 46.
b) Add 1 to x, then double. Don't forget brackets.
   The term after x is $2(x + 1)$.

■ Example 8.8
Fig. 8.2 shows a sequence of shapes made with sticks. Draw the next shape. How many sticks are needed to make the shape with n holes?

Fig. 8.2

Solution
The next shape is shown in Fig. 8.3.

Fig. 8.3

The number of sticks in a shape forms the sequence 4, 7, 10, 13, ... This sequence increases by 3, hence the nth term must involve $3n$.

52 ALGEBRAIC EXPRESSIONS

Compare the stick shape sequence with the number sequence whose *n*th term is $3n$:

	1st	2nd	3rd	4th
Stick sequence	4	7	10	13
3*n* sequence	3	6	9	12

Note that the terms of the stick sequence are each 1 greater than the terms of the $3n$ sequence. Hence we need to add 1 to $3n$.
The shape with n holes has $3n + 1$ sticks.

Key point
- The sequence 2, 7, 12, 17, 22, ... goes up in steps of 5. The *n*th term involves $5n$, not $n + 5$. The *n*th term is $5n - 3$.

EXERCISE 8C

1. Find the next two terms of the following sequences.

   a) 1, 3, 5, 7, 9
   b) 5, 8, 11, 14, 17
   c) 17, 15, 13, 11
   d) 3, 6, 12, 24, 48
   e) 2, 1, $\frac{1}{2}$, $\frac{1}{4}$, $\frac{1}{8}$
   f) 1, 1, 2, 4, 7

2. In the sequence 1, 3, 7, 15, each term is found by doubling the previous term and adding 1. Find the next two terms. Find the term after *x*.

3. In the sequence 4, 6, 12, 30, each term is found from the one before by subtracting 2 and tripling. Find the next two terms. Find the term after *x*.

4. For each of the following sequences, find an expression for the *n*th term.

   a) 1, 3, 5, 7, ...
   b) 6, 11, 16, 21, ...
   c) 1, 4, 9, 16, ...
   d) 2, 5, 10, 17, ...

5. For each of the shape sequences of Fig. 8.4, draw the next shape.

   a)

   b)

   cases

   Fig. 8.4

6. For each of the shape sequences of Fig. 8.5, find the number of dots needed to make the *n*th shape.

   a)

   b)

   Fig. 8.5

7   Milly is making a shape sequence with matchsticks, as shown in Fig. 8.6.

 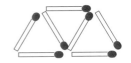 

Fig. 8.6

a) Draw the next term in the shape sequence.
b) Complete the table below, for the number of sticks in each shape.

Shape number	1	2	3	4	5
Number of sticks	3	7	11		

c) How many sticks are needed for the nth shape?
d) Which shape will need 99 sticks?

8   The *Fibonacci sequence* starts with 1, 1. Each term is then found by adding the previous two. Hence the next term is 1 + 1, i.e. 2. Write down the next three terms in the sequence.

Revision checklist

This chapter has revised:

8.1 Substitution of numbers for letters in algebraic expressions. ❑
8.2 Construction of formulae to represent quantities. ❑
8.3 Number sequences and shape sequences. Continuing sequences according to a rule. Finding the nth term of a sequence. ❑

CHAPTER 9 Equations

Chapter key points

- Apply the correct operations when solving equations. The following examples show the correct operations:

    $x + 2 = 8$ *Subtract* 2 from both sides
    $x - 2 = 8$ *Add* 2 to both sides
    $2x = 12$ *Divide* both sides by 2
    $\frac{1}{2}x = 12$ *Multiply* both sides by 2

- Do the operations in the correct order. To solve $2x + 3 = 11$, subtract 3 first, then divide by 2.
- Suppose you want to solve an equation correct to 1 decimal place. You have found that the solution lies between 1.3 and 1.4. Test at 1.35 to see whether the solution is closer to 1.3 or 1.4.
- When solving simultaneous equations, make sure you obey the basic rule for equations: do the same thing to both sides of an equation.
- Remember that subtracting a negative number is the same as adding a positive one. For example, consider the equations:

    [1] $x + 4y = 9$
    [2] $x - 5y = 3$

    Subtracting [2] from [1], we obtain $4y + 5y = 6$, not $4y - 5y = 6$.
- When solving an inequality, don't multiply or divide by a negative number. Take the negative term to the other side to make it positive.

An **equation** states that two expressions are equal. Examples of algebraic equations are:

$$2x + 3 = 9 \quad x^2 + x = 15 \quad 3x + 4y = 12$$

A letter in an algebraic equation is an **unknown**. When we find the value of the unknown, we **solve** the equation.

The basic rule for equations is:

Do to the left what you do the right

If you add 7 to the left-hand side, you must also add 7 to the right-hand side. If you divide the left by 2, you must also divide the right by 2.

9.1 Equations with one unknown

Example 9.1

Solve the equation $2x - 7 = 13$.

Solution
Add 7 to both sides:
$2x = 20$
Divide both sides by 2:
$x = 10$

Example 9.2

Solve the equation $4x + 3 = 2x + 17$. Check that your answer is correct.

Solution
Subtract $2x$ from both sides:
$2x + 3 = 17$
Subtract 3 from both sides, then divide by 2:
$2x = 14$, hence $x = 7$
Check the answer by putting $x = 7$ in both sides of the original equation:
left-hand side $= 4 \times 7 + 3 = 31$
right-hand side $= 2 \times 7 + 17 = 31$
The answer is correct.

■ Example 9.3
Solve the equation $3(x + 7) = 45$.

Solution
Note the brackets. First divide both sides by 3, then subtract 7:
$x + 7 = 15$, hence $x = 8$

Key points
■ Apply the correct operations when solving equations. The following examples show the correct operations:

$x + 2 = 8$ *Subtract* 2 from both sides
$x - 2 = 8$ *Add* 2 to both sides
$2x = 12$ *Divide* both sides by 2
$\frac{1}{2}x = 12$ *Multiply* both sides by 2

■ Do the operations in the correct order. To solve $2x + 3 = 11$, subtract 3 first, then divide by 2.

EXERCISE 9A
Solve the following equations.

1 $x + 4 = 9$
2 $y - 2 = 17$
3 $5 + x = 12$
4 $8 - x = 3$
5 $14 = 5 + x$
6 $5 = 17 - z$
7 $2x + 3 = 15$
8 $4z - 5 = 7$
9 $12 - 2a = 6$
10 $3 + 5b = 28$
11 $7 = 1 + 2x$
12 $21 = 5 + 4z$
13 $3x - 8 = 12 + 2x$
14 $3y + 5 = 11 + y$
15 $2n + 2 = 17 - 3n$
16 $5x - 4 = 2x + 11$
17 $4z + 3 = z - 12$
18 $5x + 1 = 25 - x$
19 $2(x + 3) = 10$
20 $3(b + 8) = 51$
21 $\frac{1}{4}z = 5$

9.2 Solution by trial and improvement

Some equations are harder. By **trial and improvement**, try different values until you get close to the solution. Suppose a formula is negative at $x = a$, and positive at $x = b$, then the formula will usually be 0 at a point between a and b.

Suppose you want the solution correct to 1 decimal place. Say you have found that the solution lies between 5.6 and 5.7 as shown in Fig. 9.1. Is the solution closer to 5.6 or to 5.7? Test at 5.65. In Fig. 9.1 the change between negative and positive takes place between 5.6 and 5.65, so the solution is 5.6 correct to 1 decimal place.

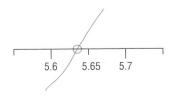

Fig. 9.1

EQUATIONS

Example 9.4

Solve the equation $x^3 + x = 4$, giving your answer correct to 1 decimal place.

Solution

First try whole number values of x:
$1^3 + 1 = 2$ too small
$2^3 + 2 = 10$ too large
So the solution lies between 1 and 2.
$1.1^3 + 1.1 = 2.431$ too small
$1.2^3 + 1.2 = 2.928$ too small
$1.3^3 + 1.3 = 3.497$ too small
$1.4^3 + 1.4 = 4.144$ too large
So the solution lies between 1.3 and 1.4. To find which it is closer to, find the value of $x^3 + x$ at 1.35:
$1.35^3 + 1.35 = 3.810$ too small
The change takes place between 1.35 and 1.4. Hence the solution is between 1.35 and 1.4.
Correct to 1 decimal place, the solution is 1.4.

> **Key point**
>
> ■ Suppose you want to solve an equation correct to 1 decimal place. You have found that the solution lies between 1.3 and 1.4. Test at 1.35, to see whether the solution is closer to 1.3 or 1.4.

> **EXERCISE 9B**
>
> 1 Solve the following by trial and improvement, giving your answers correct to 1 decimal place.
>
>   a) $x^2 = 17$ b) $x^3 = 23$ c) $x^3 = 31$
>   d) $x^2 + x = 29$ e) $x^3 + x = 7$ f) $x^3 + x = 0.8$
>   g) $x^2 + 2x = 7$ h) $x^3 + x^2 = 20$ i) $x^3 - x = 10$
>
> 2 Solve the following, giving your answers correct to 2 decimal places.
>
>   a) $x^2 + x = 3$ b) $x^3 + 2x = 2.7$ c) $x^2 - x = 0.4$

9.3 Simultaneous equations

Simultaneous equations consist of two equations with two unknowns, x and y. To solve, you need to eliminate one of the unknowns, perhaps y. Then you have an equation in x only.

Example 9.5

Solve the equations:
[1] $3x + y = 18$
[2] $2x + y = 11$
Check your answer.

Solution

Eliminate, i.e. get rid of, the y term by subtracting [2] from [1]:
[3] $x = 18 - 11 = 7$
This value of x can be put into either of the original equations to find y.

Using equation [1]:
$3 \times 7 + y = 18$, hence $y = -3$
The solutions are $x = 7$ and $y = -3$.
Equation [1] was used to find the value of x. The working can be checked by putting the values in [2]:
$2 \times 7 + -3 = 14 - 3 = 11$
The solutions are correct.

■ Example 9.6

Solve the equations:
[1] $2x + 3y = 24$
[2] $x - 4y = 1$

Solution

Multiply [2] by 2, to match the x terms:
[3] $2x - 8y = 2$
Subtract [3] from [1]. Be careful with the negative value: $3y - (-8y) = 11y$.
$11y = 22$, hence $y = 2$
Substitute in [1]:
$2x + 6 = 24$, hence $x = 9$
The solutions are $x = 9$ and $y = 2$.

■ Example 9.7

Solve the equations:
[1] $7x + 3y = 24$
[2] $5x - 2y = 13$

Solution

Here both equations must be multiplied by a number in order to match either the x terms or the y terms. Multiply [1] by 2 and [2] by 3 to match the y terms:
[3] $14x + 6y = 48$
[4] $15x - 6y = 39$
Eliminate y by adding [3] and [4]:
$29x = 87$, hence $x = 3$
Substitute in either [1] or [2] to find that $y = 1$.
The solutions are $x = 3$ and $y = 1$.

Key points

- Make sure you obey the basic rule for equations: do the same thing to both sides of an equation. For example, go from:

    $2x + 3y = 4$

    to:

    $4x + 6y = 8$

    by multiplying left and right by 2.

- Remember that subtracting a negative number is the same as adding a positive one. For example, consider the equations:

    [1] $x + 4y = 9$
    [2] $x - 5y = 3$

    Subtracting [2] from [1], we obtain $4y + 5y = 6$, not $4y - 5y = 6$.

EXERCISE 9C

Solve the following pairs of simultaneous equations.

1. $x + y = 3$
   $2x + y = 7$

2. $2x + y = 8$
   $3x + y = 10$

3. $x + 5y = 11$
   $x + 2y = 8$

4. $x + y = 17$
   $x - y = 5$

5. $3x + 2y = 6$
   $5x - 2y = 10$

6. $2p + 3q = 17$
   $-2p + 5q = 7$

7. $3x + y = 18$
   $3x - 4y = 3$

8. $2x + 3y = 17$
   $2x - 4y = 10$

9. $2x + 3y = 47$
   $x + y = 20$

10. $42m + 31n = 387$
    $m + n = 10$

11. $3x + 7y = 26$
    $x + 3y = 10$

12. $5x + 9y = 24$
    $2x + 3y = 9$

13. $4q + 3p = 39$
    $q + p = 11$

14. $3x - 2y = 6$
    $x + y = 7$

15. $5x + 7y = 13$
    $2x + 3y = 8$

16. $3z + 5w = 38$
    $2z + 3w = 23$

17. $2x + 3y = 16$
    $9x + 4y = 53$

18. $4q - 3p = 5$
    $7q - 6p = 5$

19. $3x + 2y = 14$
    $x - 3y = 1$

20. $2x - 4y = 14$
    $3x + y = 7$

21. $3x + 2y = 20$
    $-5x + 3y = 11$

9.4 Solving problems with equations

You can often solve a problem with an equation.

■ Example 9.8

The rental for an electrical tool is £25 down payment, then £7 per hour. What is the charge for x hours use?
Pete is charged £60. Form an equation in x and find how long he used the tool for.

Solution
Each hour costs £7, hence x hours cost £$7x$. Add this to the down payment.
The cost is £$(25 + 7x)$.
Put this equal to £60, obtaining the equation:
$25 + 7x = 60$
Subtract 25 from each side, obtaining $7x = 35$. Hence $x = 5$.
The tool was used for 5 hours.

■ Example 9.9

Two packets of butter and three loaves cost £4.15, while three packets of butter and five loaves cost £6.60. Let the cost of one packet of butter be xp, and the cost of one loaf of bread be yp. Form two equations in x and y, and find the cost of a loaf of bread.

Solution
There are two pieces of information, which lead to the following simultaneous equations. Write all the money in pence.
$2x + 3y = 415$
$3x + 5y = 660$
These can be solved to give $y = 75$ (and $x = 95$).
One loaf costs 75p.

EXERCISE 9D

1
  a) A plumber has a call-out charge of £15, and then he charges £30 per hour. How much does he charge for a job if it takes him x hours?
  b) If the total charge is £105 pounds, write down an equation in x. How long has the job taken him?

2
  a) To hire a car I need to pay £30, and £0.10 for each mile travelled. If I cover x miles, how much will the charge be?
  b) The total hire charge comes to £48. Write down an equation in x, and find how far I travelled.

3
  a) Mary tells Nigel: 'Think of a number, double it and add 7'. If Nigel thinks of x, what number does he end with?
  b) Nigel's final number is 29. What number had he thought of?

4
  a) George weighs 5 kg more than Edward. If Edward weighs x kg, what is their total weight?
  b) If their total weight is 149 kg, write down an equation in x. How much does Edward weigh?

5 The sum of the angles in a triangle is 180°. The three angles are $x°$, $x° + 20°$, and $x° - 5°$. Find x.

6 I buy x stamps at 20p each, and $x + 2$ stamps at 26p each. If I pay 604p in total, write down an equation in x. Solve this equation.

7 A whole number is n. What is the next number? The sum of these numbers is 93. Write down an equation in n, and solve this equation.

8 In Colin's job, the hourly rate for overtime is £4 greater than the hourly rate for ordinary time. If the ordinary rate is £x, write down an expression for the overtime rate. Colin works 38 hours ordinary time and 7 hours overtime. He receives £298. Write down an equation in x. Solve this equation.

9 3 kg of apples and 4 kg of pears cost £4.18. 1 kg of apples and 3 kg of pears cost £2.36. If 1 kg of apples and 1 kg of pears cost xp and yp respectively, write down two equations in x and y. Solve these equations to find the cost of 1 kg of apples.

10 Rick has £x and Sue has £y. Rick has £557 more than Sue. If they put their money together they have £973. How much do they each have?

11 At a concert the cheap tickets cost £5 and the more expensive ones cost £9. A total of 1000 tickets were sold, and the total receipts were £6440. Suppose x cheap tickets and y expensive tickets were sold. Write down two equations in x and y. How many expensive tickets were sold?

12 A money changer has a bundle of notes, which are either 200 franc notes or 500 franc notes. The total value of the bundle is 32 600 francs, and there are 100 notes in total. How many 500 franc notes does she have?

9.5 Inequalities

An **inequality** uses one of the following symbols:

$$< \quad > \quad \leq \quad \geq$$

$x < y$ means that x is less than y
$x > y$ means that x is greater than y
$x \leq y$ means that x is less than or equal to y (or x is at most y)
$x \geq y$ means that x is greater than or equal to y (or x is at least y)

There is an extra principle when dealing with an inequality:

Do not multiply or divide by a negative number

For example, multiplying the inequality $2 > -5$ by -1 gives $-2 > 5$, which is wrong.

■ Example 9.10

Solve the inequalities
a) $4x + 3 < 27$　　b) $4 - 3x \geq 10 - 5x$

Solution
a) Subtract 3 from both sides, obtaining $4x < 24$. As 4 is a positive number, we can divided by it.
The solution is $x < 6$.
b) We want the coefficient of x to be positive. Add $5x$ to both sides:
$4 + 2x \geq 10$
Subtract 4 and divide by 2.
The solution is $x \geq 3$.

Key point

■ When solving an inequality, do not multiply or divide by a negative number. Take the negative term to the other side to make it positive.

EXERCISE 9E

1. Solve the following inequalities.

    a) $2x + 1 < 7$　　　　b) $3x - 8 > 13$　　　　c) $4x - 3 \leq 9$
    d) $\frac{1}{3}x + 3 < 8$　　　e) $\frac{1}{4}x - 1 > 5$　　　f) $7 - x < 1$
    g) $3x - 1 \leq x + 9$　h) $2x + 3 > 18 - 3x$　i) $3 - 4x \leq 11 - 6x$

2. I buy x cakes at 40p each, and one bun at 30p. If I spend less than £5, write down an inequality in x. Solve the inequality.

3. I buy 10 stamps at xp each, and 20 stamps at $(x + 4)$p each. If I spend over £10, form an inequality in x. Solve the inequality.

Revision checklist

This chapter has revised:

9.1 Solution of simple equations with one unknown. ❏
9.2 Solution of more complicated equations, by the method of trial and improvement. ❏
9.3 Solution of simultaneous equations, in both x and y. ❏
9.4 Using equations to solve problems. ❏
9.5 Solution of inequalities. ❏

CHAPTER 10 Algebraic manipulation

Chapter key points

- When multiplying pairs of brackets in expanding an algebraic expression, *both* terms in the first pair must multiply *both* terms in the second pair. Don't just multiply the first two and the last two:

  $(a + b)(x + y) \neq ax + by$

  It should be $ax + ay + bx + by$.

- Be careful when squaring:

  $(x + y)^2 \neq x^2 + y^2$

  It should be $x^2 + 2xy + y^2$.

- Be careful with negative terms. When $(x - y)(a - b)$ is expanded, the term yb is positive.

- When simplifying an algebraic expression, combine like terms – those with the same letter to the same power:

  $5x^2 + 3x^2 = 8x^2$

- When factorising an algebraic expression, do so as fully as you can:

  $2xy + 4xz = x(2y + 4z) = 2x(y + 2z)$

- Be careful with signs when factorising:

  $x^2 - 5x + 6 = (x - 3)(x - 2)$ not $(x + 3)(x + 2)$
  $x^2 - 3x - 10 = (x - 5)(x + 2)$ not $(x + 5)(x - 2)$

- Be careful with signs when solving an equation. If $(x - 3)(x + 2) = 0$, then the solutions are $x = +3$ or -2, not $x = -3$ or $+2$.

- Do the right operation when changing the subject of a formula. If a term is *added* to the new subject, you *subtract* it from each side. If a term is *multiplying* the new subject, you *divide* each side by it.

- Do the operations in the correct order. To make x the subject of $y = 2x + 3$, first subtract 3 from each side and then divide by 2.

There are methods and rules for manipulating algebraic expressions, to get them in the most convenient form.

10.1 Expansion and simplification

Letters in algebraic expressions obey the same rules as numbers. Multiplication and division are done before addition and subtraction. If addition or subtraction need to be done first, use brackets.

    $ax + y$ means multiply a and x, then add y
    $a(x + y)$ means add x and y, then multiply by a

Expansion

To get rid of brackets, **expand**. Multiply all the terms inside the brackets by the term outside the brackets.

ALGEBRAIC MANIPULATION

$$a(x + y) = ax + ay$$

With two pairs of brackets, all the terms in the first pair multiply all the terms in the second pair.

$$(a + b)(x + y) = ax + ay + bx + by$$

A special case of this expansion is the square of an expression.

$$(x + y)^2 = (x + y)(x + y)$$
$$= x^2 + xy + yx + y^2 = x^2 + 2xy + y^2$$

Simplification

Sometimes you can simplify an expression. Terms with the same letters are **like terms**. They can be added or subtracted.

$7x$ and $5x$ are like terms: $7x + 5x = 12x$
$7x$ and $5y$ are not like terms: $7x + 5y$ cannot be simplified

Like terms must have the same power. $3x^2$ and $5x^2$ are like terms.

$3x^2 + 5x^2 = 8x^2$
$3x + 5x^2$ cannot be simplified in this way

■ Example 10.1

Expand and simplify $5(x + 3y) + 2(x - 3y)$.

Solution
Expand both pairs of brackets.
$5(x + 3y) + 2(x - 3y) = 5x + 15y + 2x - 6y$
Simplify by collecting like terms, i.e. collect the x terms and collect the y terms.
$5(x + 3y) + 2(x - 3y) = 7x + 9y$

■ Example 10.2

Expand and simplify $(2a + 3)(3a - 1)$.

Solution
Both terms in the first pair multiply both terms in the second pair.

$$(2a + 3)(3a - 1) = 6a^2 - 2a + 9a - 3$$

The $2a$ and the $9a$ are like terms.
$(2a + 3)(3a - 1) = 6a^2 + 7a - 3$

Key points

- When multiplying pairs of brackets, *both* terms in the first pair must multiply *both* terms in the second pair. Don't just multiply the first two and the last two:

    $(a + b)(x + y) \neq ax + by$

    It should be $ax + ay + bx + by$.
- Be careful when squaring:

    $(x + y)^2 \neq x^2 + y^2$

    It should be $x^2 + 2xy + y^2$.

- Be careful with negative terms. When $(x - y)(a - b)$ is expanded, the term yb is positive.
- Combine like terms – those with the same letter to the same power:

$$5x^2 + 3x^2 = 8x^2$$

but $5x^2 + 3x + 4y$ cannot be simplified.

EXERCISE 10A

Expand the following and simplify as far as possible.

1. $3(x + 8)$
2. $5(p + q)$
3. $4(3a + b)$
4. $5(4x - 7y)$
5. $2(x + 3) + 3(x - 5)$
6. $8(p + 2) - 6(1 + p)$
7. $3(p + q) + 2(4p + 2q)$
8. $2(a - 3b) - 3(a + 2b)$
9. $(x + 2)(x + 3)$
10. $(z + 2)(z + 5)$
11. $(t - 3)(t + 2)$
12. $(x + 5)(x - 6)$
13. $(w - 5)(w - 6)$
14. $(y - 3)(y - 4)$
15. $(x + y)(x - y)$
16. $(3x + 2y)(2x - 3y)$
17. $(4a - 3b)(5a - 3b)$
18. $(x + 2)^2$
19. $(3y - 2)^2$
20. $(a + b)(c + d)$
21. $(x - y)(z - w)$

10.2 Factorisation

The opposite of expansion is **factorisation**. You factorise an expression by writing it as the product of simpler expressions.

Example 10.3

Factorise:

a) $4ab - 6ac$ b) $y^2 + 2y$

Solution

a) Look for the common factors of $4ab$ and $6ac$. Both terms are divisible by a and by 2.
$4ab - 6ac = 2a(2b - 3c)$

b) Both terms are divisible by y.
$y^2 + 2y = y(y + 2)$

Key point

- Factorise as fully as you can:

$$2xy + 4xz = x(2y + 4z) = 2x(y + 2z)$$

Don't stop at the middle stage.

EXERCISE 10B

Factorise the following as far as possible.

1. $ab + ac$
2. $xy - 3y$
3. $2a + ab$
4. $3ab - 5ac$
5. $4xy + 2zy$
6. $6ab + 4ac$
7. $x^2 + 3x$
8. $y + y^2$
9. $3x^2 - 6x$
10. $15z^2 - 12z$
11. $9a^2 + 3ab$
12. $6p + 24p^2q$

ALGEBRAIC MANIPULATION

Factorising $x^2 + ax + b$

Sometimes you can factorise $x^2 + ax + b$ as $(x + \alpha)(x + \beta)$. Here α and β are numbers whose product is b and whose sum is a:

$$(x + \alpha)(x + \beta) = x^2 + (\alpha + \beta)x + \alpha\beta$$

So look for factors of b which add up to a, *or*, if b is negative, look for factors whose *difference* is a.

Solving equations

Equations of the type $x^2 + ax + b = 0$ can be solved easily if the expression $x^2 + ax + b$ can be factorised as $(x + \alpha)(x + \beta)$. If $(x + \alpha)(x + \beta) = 0$, then $(x + \alpha) = 0$ or $(x + \beta) = 0$. Hence $x = -\alpha$ or $x = -\beta$. For example, if $(x + 4)(x - 6) = 0$, then $x = -4$ or $x = 6$.

■ Example 10.4

Factorise:
a) $x^2 + 7x + 12$ b) $x^2 + 3x - 10$

Solution

a) Write out the factors of 12 and find their sum:
12 and 1, sum = 13
6 and 2, sum = 8
4 and 3, sum = 7, so these are the values of α and β.
$x^2 + 7x + 12 = (x + 4)(x + 3)$

b) Write out the factors of 10 and find their difference:
10 and 1, difference = 9
5 and 2, difference = 3, so these are the values of α and β.
As the $3x$ term is positive, take $+5$ and -2.
$x^2 + 3x - 10 = (x + 5)(x - 2)$

■ Example 10.5

Solve the equation $x^2 - 5x + 6 = 0$.

Solution

The left-hand side factorises to $(x - 3)(x - 2)$. Note the minus signs.
Hence $x = 3$ or $x = 2$.

Key points
■ Be careful with signs when factorising:

$$x^2 - 5x + 6 = (x - 3)(x - 2) \quad \text{not} \quad (x + 3)(x + 2)$$

$$x^2 - 3x - 10 = (x - 5)(x + 2) \quad \text{not} \quad (x + 5)(x - 2)$$

■ Be careful with signs when solving an equation. If $(x - 3)(x + 2) = 0$, then the solutions are $x = +3$ or -2, not $x = -3$ or $+2$.

EXERCISE 10C

1 Factorise the following expressions.

a) $x^2 + 5x + 6$ b) $x^2 + 3x + 2$ c) $x^2 + 2x + 1$
d) $x^2 - 4x + 3$ e) $x^2 - 6x + 8$ f) $x^2 + 4x - 5$
g) $x^2 + 2x - 8$ h) $x^2 - 3x - 18$ i) $x^2 - x - 20$

2 Solve the following equations.

a) $x^2 + 4x + 3 = 0$
b) $x^2 + 8x + 15 = 0$
c) $x^2 - 6x + 5 = 0$
d) $x^2 - 7x + 10 = 0$
e) $x^2 + 4x - 5 = 0$
f) $x^2 + 2x - 15 = 0$
g) $x^2 - 2x - 3 = 0$
h) $x^2 - x - 12 = 0$

10.3 Changing the subject

In the formula $y = 3x + 2$, y is expressed in terms of x. The **subject** of the formula is y. If we rewrite the formula so that x is expressed in terms of y, then we have changed the subject of the formula.

■ Example 10.6

If $y = 2x + c$, put x in terms of y and c.

Solution
Subtract c from both sides, to obtain $y - c = 2x$. Then divide by 2.
$x = \frac{1}{2}y - \frac{1}{2}c = \frac{1}{2}(y - c)$

■ Example 10.7

The formula $C = (F - 32) \times \frac{5}{9}$ converts temperature in Fahrenheit to Celsius. Rearrange the formula to express F in terms of C, i.e. to convert Celsius to Fahrenheit.

Solution
Multiply both sides by 9, to get $9C = (F - 32) \times 5$.
Divide both sides by 5, to get $\frac{9C}{5} = F - 32$. Then add 32.
$F = \frac{9C}{5} + 32$

EXERCISE 10D

1 For each of the following formulae, change the subject to the letter in brackets.

a) $y = 3x + 2$ (x)
b) $k = 5m - 10$ (m)
c) $z = 5w + 3$ (w)
d) $y = w + x - z$ (z)
e) $3x + 2y = 4$ (x)
f) $7y = 8 - x$ (x)
g) $y = \frac{1}{3}x - 4$ (x)
h) $y = \frac{1}{4}x + 3$ (x)
i) $y = \frac{2}{3}z - 8$ (z)
j) $ab = cd$ (c)
k) $y = 4ax$ (x)
l) $b = \frac{1}{2}(c + a)$ (a)
m) $y = mx + c$ (x)
n) $y + 5 = \frac{1}{2}xz$ (x)
o) $\frac{7}{x} = 3y$ (y)

2 The formula $F = \frac{9}{4}R + 32$ converts temperature on the Réaumur scale to temperature in Fahrenheit. Rearrange the formula so that it converts Fahrenheit to Réaumur.

3 The velocity v of a body is given by $v = u + at$. Rearrange the formula so that a is the subject.

4 The volume V of a cylinder is given by $V = \pi r^2 h$. Rearrange the formula so that h is the subject.

Key points

■ Do the right operation when changing the subject of a formula:
  - If a term is *added* to the new subject, you *subtract* it from each side.
  - If a term is *multiplying* the new subject, *you divide* each side by it.

■ Do the operations in the correct order. To make x the subject of $y = 2x + 3$, first subtract 3 from each side and then divide by 2.

Revision checklist

This chapter has revised:

10.1 Expansion of expressions by removing brackets. Simplification of expressions by collecting like terms. ❏
10.2 Factorisation of expressions by taking out common factors. Factorising $x^2 + ax + b$ and solving equations of the form $x^2 + ax + b = 0$. ❏
10.3 Rearranging a formula to change its subject. ❏

CHAPTER 11 Coordinates and graphs

Chapter key points

- The *x*-coordinate is always given first. The point at (3, 2) is at $x = 3$ and $y = 2$. Be sure to get this the right way round.
- Make sure that the scale of your graph is uniform. It is misleading to write down a scale in which the *x*-values or *y*-values are not evenly spaced.
- Be careful when reading or plotting fractional coordinates. Count how many divisions there are between whole numbers.
- If a graph converts between currencies, £ and $ say, make sure that you convert in the right direction.

11.1 Coordinates

We represent the position of a point by **coordinates** relative to two lines. The vertical line is the ***y*-axis**, and the horizontal line is the ***x*-axis**. The position relative to these lines is given by the ***x*-coordinate** and the ***y*-coordinate**.

In Fig. 11.1, the point A has *x*-coordinate 3 and *y*-coordinate 2. Hence it is at (3, 2). Note that the *x*-coordinate is given first.

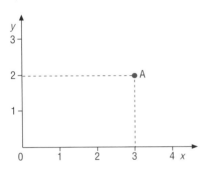

Fig. 11.1   Fig. 11.2

Coordinates can be negative. In Fig. 11.2 the point B is at $(-2, -1)$.

The axes cross at the **origin**. At the origin O, $x = y = 0$. Hence the origin is at (0, 0).

■ Example 11.1

On the graph of Fig. 11.3 mark the point Z with coordinates (2.6, 1.3).

Solution

Note that the graph grid has 5 divisions between whole numbers. Hence each division corresponds to $\frac{1}{5}$, i.e. 0.2. So the *x*-coordinate of Z is 3 divisions to the right of 2, and its *y*-coordinate is $1\frac{1}{2}$ divisions above 1.

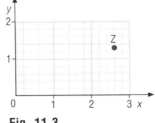

Fig. 11.3

Key points

- The *x*-coordinate is always given first. The point at (3, 2) is at $x = 3$ and $y = 2$. Be sure to get this the right way round.
- Make sure that the scale of your graph is uniform. It is misleading to write down a scale in which the *x*-values or *y*-values are not evenly spaced.
- Be careful when reading or plotting fractional coordinates. Count how many divisions there are between whole numbers.

EXERCISE 11A

1. Fig. 11.4 shows a map of part of a park. The pond is in square C5.

   a) Write down the square containing the tennis courts.
   b) Elm Wood is in more than one square. Write down these squares.

2. In Fig. 11.5 there are points A, B and C. Write down their coordinates.

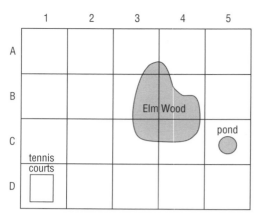

Fig. 11.4

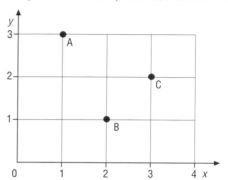

Fig. 11.5

3. On the graph of Fig. 11.5 mark the points L(1, 1), M(2, 3), N(4, 1).

4. Write down the coordinates of the points X, Y, Z in Fig. 11.6.

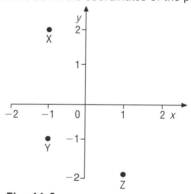

Fig. 11.6

Fig. 11.7

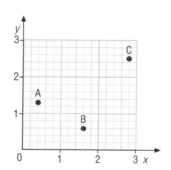
Fig. 11.8

5. On the graph of Fig. 11.7 mark the points A(1, 1), B(1, 3), C(4, 3).

   a) ABCD is a rectangle. Mark D on the graph, and write down its coordinates.
   b) What are the coordinates of the midpoint of AC?

6. Write down the coordinates of the points A, B, and C in Fig. 11.8.

7. On the graph of Fig. 11.8 plot the points with the following coordinates:

   a) (0.8, 2.2)   b) (1.8, 2.9)   c) (2.5, 1.3)

8. On the graph of Fig. 11.9 plot the points with the following coordinates:

   a) (14, 8)   b) (15, 15)   c) (24, 27)

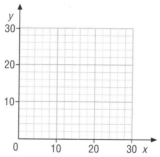
Fig. 11.9

9. The squares on a chess board are identified by a letter and a number. In Fig. 11.10, the square marked * is e4.

   a) What is the square marked **?
   b) The King can move one square in any direction. Where can he move to from the square marked *?
   c) The Knight moves in an L shape: two spaces up or down, followed by one space to the right or left, or two spaces right or left, followed by one space up or down. Where can he move to from the square marked *?

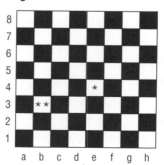
Fig. 11.10

10 A cell in a spreadsheet is identified by the letter of its column and the number of its row. See Fig. 11.11.

a) What cell is immediately below C5?
b) What cell is immediately to the right of E7?
c) From cell N12, go up 3 and 3 to the left. Where are you now?

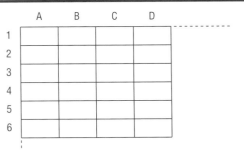

Fig. 11.11

11.2 Interpretation of graphs

If two quantities are connected, the connection can be shown by a graph.

■ Example 11.2

During a heat wave the temperature was measured every 4 hours. The results are in the table below.

Time	1200	1600	2000	2400	0400	0800	1200
Temp. °F	88	91	85	64	62	63	84

Plot these figures on a graph of temperature against time. From your graph estimate:
a) the highest temperature
b) the time for which the temperature was under 70°F.

Solution
The graph of temperature against time is shown in Fig. 11.12.

a) Read off the highest value.
   The highest temperature was 92°F.
b) The temperature was under 70°F between 2200 and 1000.
   The temperature was under 70°F for 12 hours.

■ Example 11.3

Different weights are suspended from a string. The length of the wire is l cm, and the weight suspended is W kg. l and W obey the following law:

$$l = 10 + 8W$$

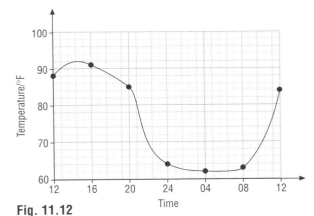

Fig. 11.12

Find the lengths of the string when the weights are 1 kg, 2 kg and 3 kg.
  Plot your results on a graph, using the weights for the x-values and the lengths for the y-values.

Solution

Apply the formula to the values.
For $W = 1, l = 10 + 8 \times 1 = 18$
For $W = 2, l = 10 + 8 \times 2 = 26$
For $W = 3, l = 10 + 8 \times 3 = 34$
Plot the points (1, 18), (2, 26), (3, 34) on a graph (Fig. 11.13). Join them up by a straight line.

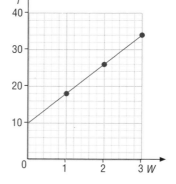

Fig. 11.13

Key point

■ If a graph converts between currencies, £ and $ say, make sure that you convert in the right direction.

EXERCISE 11B

1. The graph of Fig. 11.14 shows the conversion rate between £ and FF (French francs), at a certain time.

   a) How much is £2 in FF?
   b) How much is 26FF in £?

   After a while the rate changes so that £1 is worth 11FF. Draw a line on the graph for the new exchange rate.

2. At a certain time there are 2.5 German marks (DM) to the £. Draw a line on Fig. 11.15 for conversion between £ and DM. Use your graph to:

   a) convert £18 to DM   b) convert 12DM to £

3. In Fig. 11.16 the solid line converts from £ to $. The broken line converts back from $ to £.

   a) I change £24. How many dollars do I get?
   b) I now change these dollars back to £. How much do I get?

4. A nation's unemployment rate, as a percentage, is recorded over a period of six years. Fig. 11.17 shows the graph of the unemployment rate.

   a) What was the highest level of unemployment?
   b) For how long was unemployment over 10%?

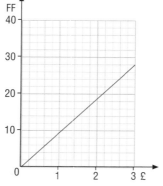

Fig. 11.14

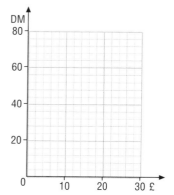

Fig. 11.15

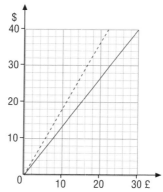

Fig. 11.16

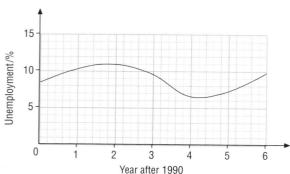

Fig. 11.17

5. Leonora has a fever. Her temperature over a period of 24 hours is shown in the graph of Fig. 11.18.

   a) What were her greatest and her least temperatures?
   b) When was her temperature 98°F?

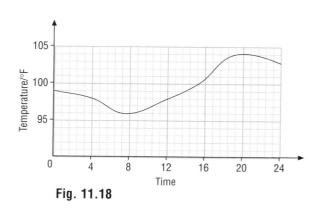

Fig. 11.18

6 The government of a country keeps a record of inflation over a period of eight years. The results are given by the following table.

Year (31 December)	1987	1988	1989	1990	1991	1992	1993	1994
Inflation (%)	12	16	23	20	18	14	9	8

Draw a graph to illustrate these figures, taking 1 cm along the *x*-axis to represent 1 year, and 1 cm up the *y*-axis to represent 5%.

a) What was the highest level of inflation?
b) When did inflation start to come down?

7 A gun is fired at different angles of elevation. The range of the gun is measured, as shown in the following table.

Angle of elevation	10°	20°	30°	40°	50°	60°
Range in metres	1370	2600	3460	3940	3920	3400

Plot these figures, taking 1 cm along the *x*-axis to represent 10°, and 1 cm up the *y*-axis to represent 500 m. Join up the dots as smoothly as you can.

a) What was the greatest range? What angle gives the greatest range?
b) What angles give a range of 3500 m?

8 The Acme Van Hire company charges £12 plus £2 per mile. The Nadir company charges a flat rate of £3 per mile. Complete the following table for the cost of hiring.

Miles	5	10	15	20	25
Acme charge (£)	22				
Nadir charge (£)	15				

Plot the points on the graph of Fig. 11.19. Join up by straight lines. When is it cheaper to use Acme rather than Nadir?

9 The rule for converting temperature in Fahrenheit, $F°$, to temperature in Réaumur, $R°$, is:

$$R = \tfrac{4}{9}(F - 32)$$

Find the Réaumur temperatures corresponding to 0°F, 50°F and 100°F. Construct a chart to convert from Fahrenheit to Réaumur.

10 Water is being poured at a steady rate into a conical flask. Which of the pictures of Fig. 11.20 could represent the graph of the water level against time?

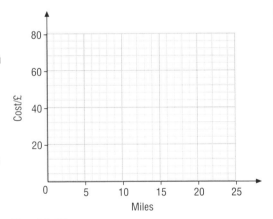

Fig. 11.19

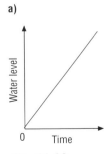
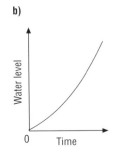
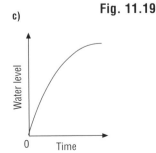

Fig. 11.20

11.3 Travel graphs

A graph of distance against time is a **travel graph**. It describes a journey. The speed of a stage of the journey can be found from the graph.

■ Example 11.4

As part of her training, Donna runs a distance, then pauses for a rest, then runs back. Fig. 11.21 is a distance–time graph showing her distance from the starting point.

a) How far did she run in the first stage?
b) For how long did she rest?
c) What was her speed in the final stage?

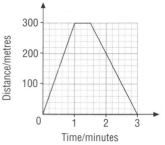

Fig. 11.21

Solution

a) The total distance up the *y*-axis is 300 m.
   She ran 300 m in the first stage.
b) The rest stage occupies $\frac{1}{2}$ minute of the *x*-axis.
   She rested for $\frac{1}{2}$ minute.
c) In the final stage she ran 300 m in $1\frac{1}{2}$ minutes. Divide 300 by $1\frac{1}{2}$, obtaining 200.
   Her speed was 200 metres per minute.

EXERCISE 11C

1. Piers swims across a lake. Part of the way across there is a raft on which he rests. Fig. 11.22 is a distance–time graph showing his distance across the lake.

   a) How wide is the lake?
   b) For how long did Piers rest?
   c) What was his speed in the final stage of his swim?

2. Patti drives to the library to return a book. On the way she realises that she has forgotten to bring the book and goes home to collect it. Fig. 11.23 is a graph of her distance from home. She reaches the library after 15 minutes.

   a) After how long did she turn back?
   b) How far away is the library?

3. Dennis walks to the station, waits for a train, then takes the train to his workplace. Fig. 11.24 shows a graph of his distance from home.

   a) How far is the station from his home?
   b) How long did the train take between the stations?
   c) What was the speed of the train?

4. A lift rises from the ground floor of a store to the second floor, then descends. Fig. 11.25 shows its height above the ground floor.

   a) How long did it stay at the first and second floors?
   b) How high is the second floor above the ground floor?
   c) How fast is the lift when it is descending?

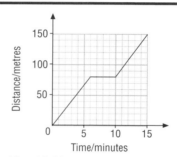

Fig. 11.22

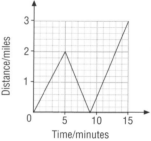

Fig. 11.23

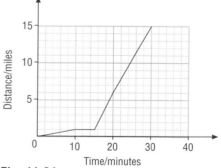

Fig. 11.24

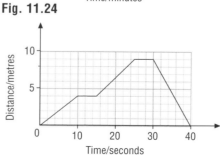

Fig. 11.25

5   Robert sets off from home to a shop, walking 3 miles in 1 hour. He stays in the shop for $\frac{1}{2}$ hour, then walks home in $1\frac{1}{2}$ hours. On Fig. 11.26 draw a travel graph to show Robert's journey.

6   Jo lives 400 m due north of Dawn. They set out to visit each other at the same time; Jo walks south at 100 m per minute and Dawn walks north at 50 m per minute. On the diagram of Fig. 11.27 draw graphs showing their distances from Dawn's house. When do they meet? How far north of Dawn's house do they meet?

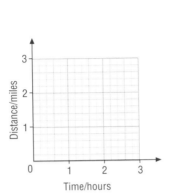

Fig. 11.26

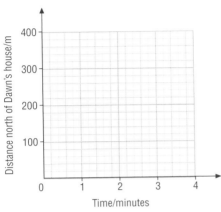

Fig. 11.27

Revision checklist

This chapter has revised:

11.1 The use of coordinates to represent position.   ❏
11.2 The use of a graph to show the connection between two quantities.   ❏
11.3 The use of a travel graph to describe a journey.   ❏

CHAPTER 12 Functions and graphs

Chapter key points

- Be careful when working out tables of values of functions, particularly when negative numbers are involved. When $x = -1$, the value of $2 - x^2$ is 1, not 3.
- A graph like $y = x^2$ is curved. Join the points of the graph with a smooth curve, not straight lines.
- With a complicated graph scale, such as 1 cm per 5 m, be very careful when plotting points.
- Be careful with reciprocal graphs, when fractions or 0 are involved.
  - When $x = \frac{1}{2}$, the value of $\frac{1}{x}$ is 2, not $\frac{1}{2}$.
  - When $x = 0$, $1 + \frac{1}{x}$ has no meaning.
- When solving something like $x^2 - x < 0$, don't divide by x (it may be negative). Plot the graph and find where the curve is below the x-axis.
- If you are solving simultaneous equations by the graph method, give both the x-value and the y-value. But if you are solving an equation in x by itself, give only the x-value, not the y-value.
- When solving two-dimensional inequalities, if you are not sure which side of a line to shade, take a test point on one side of the line. If the inequality is true at the point, then that is the side to shade.

12.1 Graphs of functions

Let y be a **function** of x. For each value of x there is a corresponding value of y. The graph of y consists of the pairs (x,y), plotted on graph paper.

Linear and x^2 graphs

A **linear graph** is a straight line. Example 12.1 below gives a linear graph. The graph of $y = x^2$ is curved, as we see in Example 12.2.

■ Example 12.1

Fill in the following table for the function $y = 2x - 1$. Plot the graph of the function for these values of x.

x	0	1	2	3	4
y					

Solution

For $x = 0$, $y = 2 \times 0 - 1 = 0 - 1 = -1$
For $x = 1$, $y = 2 \times 1 - 1 = 2 - 1 = 1$
For $x = 2$, $y = 2 \times 2 - 1 = 4 - 1 = 3$
For $x = 3$, $y = 2 \times 3 - 1 = 6 - 1 = 5$
For $x = 4$, $y = 2 \times 4 - 1 = 8 - 1 = 7$

The completed table is at the top of page 75.

x	0	1	2	3	4
y	−1	1	3	5	7

Plot points at (0, 1), (1, 1), (2, 3) and so on. Join the points up. The result is shown in Fig. 12.1. Notice that it is a straight line (linear) graph.

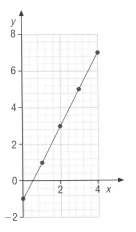

Fig. 12.1

Example 12.2
Let $y = x^2$. Find the values of y for values of x between −2 and 2. Hence plot the graph of $y = x^2$.

Solution
Fill in the table below, by squaring all the values of x. Remember the square of a negative number is positive. The square of −2 is 4, the square of −1 is 1, and so on.

x	−2	−1	0	1	2
x^2	4	1	0	1	4

Plot points at (−2, 4), (−1, 1) up to (2, 4). Join the points by a smooth curve. The result is shown in Fig. 12.2.

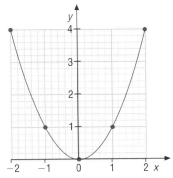

Fig. 12.2

Key points
- Be careful when working out tables of values, particularly when negative numbers are involved. When $x = -1$, the value of $2 - x^2$ is 1, not 3.
- A graph like $y = x^2$ is curved. Join the points of the graph with a smooth curve, not with straight lines.
- With a complicated scale, such as 1 cm per 5 m, be very careful when plotting points.

EXERCISE 12A

1. Let $y = 3 + x$. Find the values of y when x is 0, 1, 2, 3. Plot the graph of the function on Fig. 12.3.

2. For each of the following functions, find the values of y when x is 0, 1, 2, 3.
   Plot the functions, using a scale of 1 cm per unit.

   a) $y = 2 + x$ b) $y = 2 - x$

   c) $y = 2x + 1$

3. Let $y = x^2 + 1$. Complete the table below for the values of x and y. Plot the graph of the function on Fig. 12.4.

x	−2	−1	0	1	2
x^2	4				
$x^2 + 1$	5				

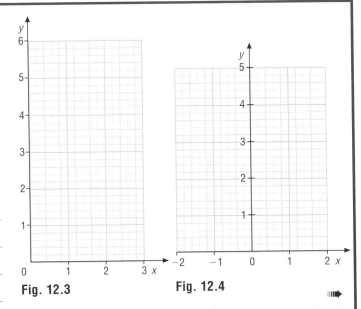

Fig. 12.3   Fig. 12.4

76 FUNCTIONS AND GRAPHS

4 Let $y = 2x^2$. Find the values of y for x equal to $-2, -1, 0, 1, 2, 3$. Plot the graph of $y = 2x^2$, using a scale of 1 cm per unit.

5 Let $y = (x - 1)^2$. Complete the table below. Plot a graph of the function, using a scale of 1 cm per unit.

x	−1	0	$\frac{1}{2}$	1	$1\frac{1}{2}$	2	3
y			$\frac{1}{4}$				

12.2 Reciprocal and quadratic graphs

A **reciprocal** function involves $\frac{1}{x}$. You cannot divide by 0, hence $\frac{1}{x}$ does not exist when $x = 0$.

A function of the form $y = x^2 + ax + b$ is a **quadratic** function. Its graph has the same shape as that of $y = x^2$. A quadratic graph can be used to solve a quadratic inequality, of the form:

$$x^2 + ax + b < 0$$

To solve the inequality, you need to find the values of x for which the graph is below the x-axis.

■ Example 12.3

y is given in terms of x by the formula $y = 2 - \frac{1}{x}$. Complete the following table.

x	1	2	3	4	5
y					

Draw a graph of the function, using a scale of 1 cm per unit.

Solution

For $x = 1$, $y = 2 - \frac{1}{1} = 2 - 1 = 1$

For $x = 2$, $y = 2 - \frac{1}{2} = 2 - \frac{1}{2} = 1\frac{1}{2}$

The other values are found similarly. The completed table is:

x	1	2	3	4	5
y	1	$1\frac{1}{2}$	$1\frac{2}{3}$	$1\frac{3}{4}$	1.8

Plot points at (1, 1), (2, 1.5) and so on. The graph is shown in Fig. 12.5.

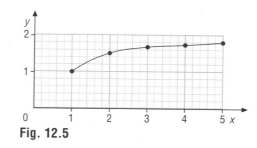

Fig. 12.5

FUNCTIONS AND GRAPHS

■ Example 12.4

Plot the graph of $y = x^2 - 4x + 3$, for values of x between -1 and 4.

Solution
Set out a table of values as shown. It helps to have separate rows for x^2 and $4x$. To find each y value, subtract $4x$ from x^2 and then add 3. For example, for $x = 2$, $y = 4 - 8 + 3 = -1$.

x	-1	0	1	2	3	4
x²	1	0	1	4	9	16
4x	-4	0	4	8	12	16
y	8	3	0	-1	0	3

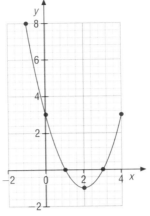

Fig. 12.6

Plot points at $(-1, 8)$, $(0, 3)$, and so on. The graph is plotted in Fig. 12.6.

■ Example 12.5

Use the graph of Example 12.4 to solve the inequality $x^2 - 4x + 3 < 0$.

Solution
The graph of $y = x^2 - 4x + 3$ is shown in Fig. 12.6. Note that the graph crosses the x-axis at $x = 1$ and $x = 3$. Between these values the curve lies below the axis, i.e. y is negative.
The solution is $1 < x < 3$.

Key points
■ Be careful with reciprocal graphs, when fractions or 0 are involved.
- When $x = \frac{1}{2}$, the value of $\frac{1}{x}$ is 2, not $\frac{1}{2}$.
- When $x = 0$, $1 + \frac{1}{x}$ has no meaning.

■ When solving something like $x^2 - x < 0$, don't divide by x (it may be negative). Plot the graph and find where the curve is below the x-axis.

EXERCISE 12B

1 Complete the table for the function $y = 1 + \frac{2}{x}$.

x	1	2	3	4	5
y					

- Draw a graph of the function on Fig. 12.7.

2 Find the value of the function $y = 3 - \frac{2}{x}$, for $x = 1, 2, 3,$ and 4. Plot the graph of the function.

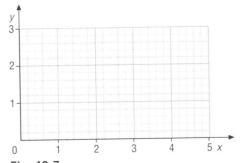

Fig. 12.7

3. Complete the table below for the function $y = x^2 - 2x$. Plot the graph of the function on Fig. 12.8.

x	−1	0	1	2	3
x^2					
$2x$					
y					

4. Find the values of the function $y = x^2 - x - 2$ for values of x between −2 and 3. Plot the graph of the function.
5. Use your graph from Question 3 to solve the inequality $x^2 - 2x < 0$.
6. Use your graph from Question 4 to solve the inequality $x^2 - x - 2 < 0$.

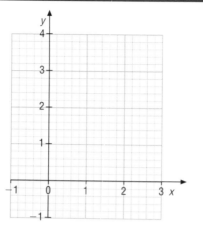

Fig. 12.8

12.3 Solving equations by graph

Simultaneous equations

Suppose you have two simultaneous equations. Plot their graphs on the same paper. The point where the graphs cross gives the solution of the simultaneous equations.

■ Example 12.6

Plot the graphs of $y = 2x - 3$ and $3x + 2y = 6$, on the same paper. Use your raphs to solve the simultaneous equations:

$$y - 2x = -3$$
$$3x + 2y = 6$$

Solution
Find points which satisfy $y = 2x - 3$.
$x = 0, y = 2 \times 0 - 3 = -3$
$x = 1, y = 2 \times 1 - 3 = -1$
$x = 2, y = 2 \times 2 - 3 = 1$

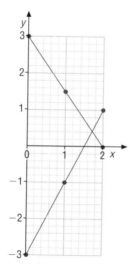

Fig. 12.09

The points are $(0, -3)$, $(1, -1)$, $(2, 1)$. They are plotted on the graph of Fig. 12.9.

Three points which satisfy $3x + 2y = 6$ are $(0, 3)$, $(1, 1\frac{1}{2})$, $(2, 0)$. They are also plotted on Fig. 12.9.

The two equations are both true at the crossing point of the two lines. Read off the coordinates of this point:
$x = 1.7, y = 0.4$

Quadratic equations

Equations of the form $x^2 + ax + b = 0$ can be solved by plotting the quadratic graph $y = x^2 + ax + b$.

■ Example 12.7

Plot the graph of $y = x^2 - 2x - 1$, for values of x between −1 and 4. Hence solve the equation $x^2 - 2x - 1 = 0$.

Solution

Write out a table of values as below.

x	−1	0	1	2	3	4
y	2	−1	−2	−1	2	7

Fig. 12.10 shows the graph. The equation is true when $y = 0$, i.e. when the curve crosses the x-axis. Read off the values of x. The solutions are $x = -0.4$ and $x = 2.4$.

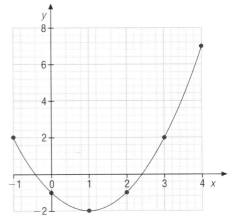

Fig. 12.10

Key point

- If you are solving simultaneous equations by the graph method, give both the x-value and the y-value. But if you are solving an equation in x by itself, give only the x-value, not the y-value.

EXERCISE 12C

1. On the graph of Fig. 12.11, plot the graphs of $y = 2x + 1$ and $2y = x + 5$. Hence solve the simultaneous equations:

   $y = 2x + 1$
   $2y = x + 5$

2. Solve the following pairs of simultaneous equations, by means of drawing lines on graph paper and finding the point of intersection.

   a) $y = x + 1$
   $y = 3x - 5$

   b) $y = 2x - 4$
   $y = 11 - 3x$

   c) $2y + 3x = 12$
   $x + 3y = 11$

3. Complete the following table for the function $y = x^2 + 3x - 1$.

x	−4	−3	−2	−1	0	1	2
y							

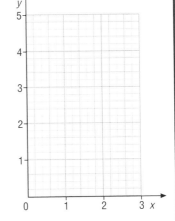

Fig. 12.11

Draw the graph of your function. Use your graph to find approximate solutions to the equation:

$x^2 + 3x - 1 = 0$

4. Complete the following table for the function $y = x + \dfrac{2}{x}$.

x	$\tfrac{1}{2}$	1	$1\tfrac{1}{2}$	2	$2\tfrac{1}{2}$	3	$3\tfrac{1}{2}$	4
y								

Draw a graph of the function. Use your graph to solve the equation:

$x + \dfrac{2}{x} = 4$

5 A cricket ball is thrown vertically upwards. After x seconds its height, y m, above the ground is given by $y = 2 + 15x - 5x^2$. Plot the graph of this function, for x between 0 and 4. Use your graph to answer the following.

   a) When does the ball reach its greatest height?
   b) What is the greatest height?
   c) When is it 5 m high?
   d) For how long is it in the air?

6 The faster a car is going, the longer it takes to stop. If a car takes a distance D feet to stop when it travels at speed s m.p.h., then D is approximately given by the formula:

   $$D = s + \frac{s^2}{20}$$

   Plot the graph of this function. Use your graph to solve the following.

   a) How long does a car travelling at 45 m.p.h. take to stop?
   b) If a car must be able to stop within 80 feet, what is the greatest speed permitted?

12.4 Two-dimensional inequalities

An inequality in x and y is a **two-dimensional inequality**. It represents a region in the two-dimensional plane.

To solve a two-dimensional inequality, plot the line corresponding to the equality. Then shade the appropriate side of the line.

■ Example 12.8

The region shaded in Fig. 12.12 is bounded by the three lines $x = \frac{1}{2}$, $y = \frac{1}{2}$ and $y + 2x = 4$. Describe the region by three inequalities.

Solution
The region lies to the right of the line $x = \frac{1}{2}$, hence $x > \frac{1}{2}$. It is above the line $y = \frac{1}{2}$, hence $y > \frac{1}{2}$. It is below $y + 2x = 4$, hence $y + 2x < 4$.
$x > \frac{1}{2}, y > \frac{1}{2}, y + 2x < 4$

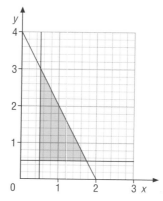
Fig. 12.12

Key point

■ If you are not sure which side of a line to shade, take a test point on one side of the line. If the inequality is true at the point, then that is the side to shade.

EXERCISE 12D

1 The shaded region of Fig. 12.13 is bounded by the lines $x = 2$, $y = 1$ and $x + y = 4$.
   Describe the region by inequalities.

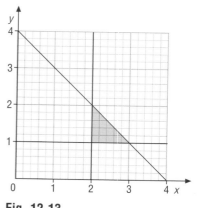
Fig. 12.13

FUNCTIONS AND GRAPHS

2 Fig. 12.14 shows the lines $x = 0$, $y = 0$, $2x + 3y = 6$, and $3x + 2y = 6$. Describe the shaded regions A and B by inequalities.

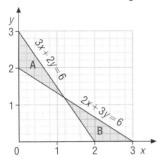
Fig. 12.14

3 In Fig. 12.15 the lines $y = 1$, $x = 1$, $y = x + 2$ and $x + y = 5$ are drawn. On the graph label regions A, B, defined by the following inequalities.

A: $x > 1$, $y > 1$, $y < x + 2$, $x + y < 5$
B: $x < 1$, $y > 1$, $y < x + 2$, $x + y < 5$

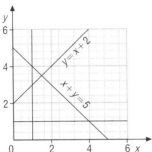
Fig. 12.15

4 Tarts cost 60p each, and scones cost 40p each. Nick has £2.40 to spend, and has been told not to buy more than 5 items. If he buys x tarts and y scones, find two inequalities in x and y. Illustrate them on the graph of Fig. 12.16, shading the region of Nick's possible choices.

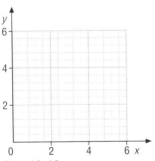
Fig. 12.16

Revision checklist

This chapter has revised:

12.1 The drawing of straight line graphs, and graphs involving x^2. ❑

12.2 Reciprocal graphs involving $\frac{1}{x}$. Quadratic graphs of the form $y = x^2 + ax + b$. Solving quadratic inequalities. ❑

12.3 Solving simultaneous equations by intersecting graphs. Solving equations by quadratic graphs. ❑

12.4 Solving two-dimensional inequalities by shading regions on graph paper. ❑

© IT IS ILLEGAL TO PHOTOCOPY THIS PAGE

Mixed exercise 2

1. Solve the following equations.

   a) $x - 7 = 8$ **b)** $4x - 5 = x + 7$ **c)** $17 - 2x = 3$

2. Given the formula $y = 3x - 7$, find y when $x = 5$.

3. **a)** Write down the coordinates of point A in Fig. M2.1.
   b) On Fig. M2.1 place point B at (1.3, 1.8).

4. Fig. M2.2 shows a sequence of patterns of dots.

   Fig. M2.2

   a) Draw the next pattern.
   b) Write down the number of dots in each of the patterns.
   c) How many dots will there be in the fifth pattern?

5. Let $y = x^2 + 2$. Plot the graph of y on Fig. M2.3. What is the least value of y?

6. Expand and simplify the expression $3(x + 7) + 2(x - 4)$.

7. A man rows his boat straight across a river. The travel graph of Fig. M2.4 shows the distance he has rowed.

   a) How wide is the river?
   b) How long did it take to cross?
   c) What is his speed of rowing?
   d) He stays on the other bank for 5 minutes, and then rows back in 8 minutes. Extend the graph of Fig. M2.4 to show his complete journey.

8. A mail order form sells shirts at £x each. I order five shirts.

   a) What is the cost of the five shirts? Give your answer in terms of x.

   There is a special offer of £7 discount for customers buying more than three shirts.

   b) How much do I have to pay for the five shirts?
   c) I have to pay £78. Write down an equation in x.
   d) Solve this equation to find x.

9. The midday temperature at a location is measured for a period of eight days. The results are given in the table below.

Day	1	2	3	4	5	6	7	8
Temperature (°C)	−5	−3	0	2	2	1	−1	−5

   a) Plot these figures on the graph of M2.5. Join them up by a smooth curve.
   b) What is the difference between the highest and the lowest temperature?
   c) Between which days was there the greatest change in temperature?

10. **a)** If $y = 3x - 7$, find y when $x = 3\frac{1}{2}$.
    b) If $z = 4x + 5y - 2$, find z when $x = 4$ and $y = 7$.

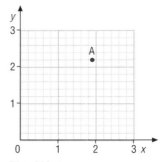

Fig. M2.1

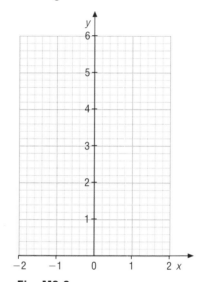

Fig. M2.3

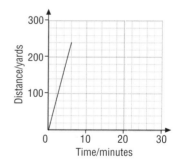

Fig. M2.4

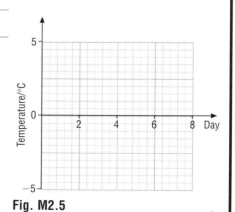

Fig. M2.5

The following questions are suitable for Intermediate level.

11 Solve the following pairs of simultaneous equations.

a) $x + y = 11$
   $x - y = 3$

b) $2x + 5y = 16$
   $3x - y = 7$

12 Look again at the patterns of dots in Fig. M2.2. How many dots will there be in the nth pattern?

13 a) Factorise the following expressions.
   i) $4ax + 14bx$ ii) $x^2 + 7x + 10$
   b) Solve the equation $x^2 - 9x + 14 = 0$.

14 On Fig. M2.6 draw the graph of $y = x^2 - 5x + 6$.

a) From your graph solve the equation $x^2 - 5x + 6 = 0$.
b) From your graph solve the inequality $x^2 - 5x + 6 < 0$.

15 Evaluate the expression below, for $t = 7$, $k = 3.2$ and $m = 4.9$. Give your answer correct to 3 significant figures.

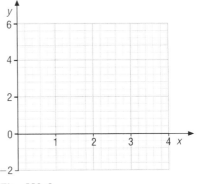

Fig. M2.6

$$t\sqrt{\left(\frac{k - 2.8}{m - 4.7}\right)}$$

16 Make v the subject of the following formulae.

a) $s = 5v + 3$ b) $T = 5av$

17 Solve the following equation, correct to 1 decimal place, by trial and improvement.

$x^3 + 2x = 7$

18 A rule for converting a temperature on the Fahrenheit scale, $F°$, to a temperature on the Celsius scale, $C°$, is:

$F = \frac{9}{5}C + 32$

At what temperature do the two scales give the same reading?

19 Solve the following inequalities.

a) $3x + 5 > 23$ b) $2x + 17 < 6x + 5$

20 Water is being poured into a container at a constant rate. Possible graphs of water level against time are shown in Fig. M2.7. Possible shapes for the container are also shown in Fig. M2.7. Match each graph with a container shape.

a)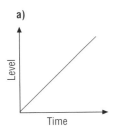

b)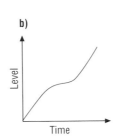

c)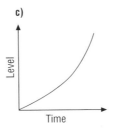

d)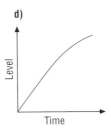

i)

ii)

iii)

iv)

Fig. M2.7

Section 3
SHAPE AND SPACE

CHAPTER 13 Plane figures

Chapter key points
- Label a polygon correctly. The letters must go either clockwise or anti-clockwise around the figure. They should not jump across a diagonal.
- Don't assume that a polygon is regular, unless you are told it is.

13.1 Angles and lines

Fig. 13.1 shows three lines meeting at a point. The angles add up to 360°.

$130° + 150° + 80° = 360°$

Fig. 13.2 shows a line meeting a straight line. The angles add up to 180°.

$80° + 100° = 180°$

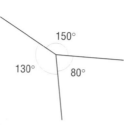

Fig. 13.1

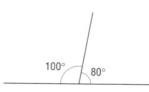

Fig. 13.2

Special angles and lines (Fig. 13.3)

Right angle	an angle of 90°.
Acute angle	an angle between 0° and 90°.
Obtuse angle	an angle between 90° and 180°.
Reflex angle	an angle greater than 180°.
Parallel lines	lines which are in the same direction.
Perpendicular lines	lines which are at right angles.

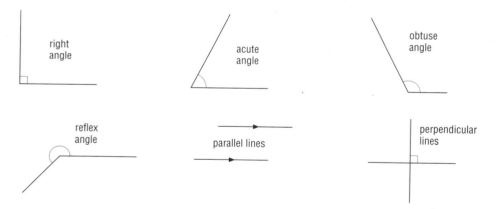

Fig. 13.3

Example 13.1

In Fig. 13.4 the horizontal and vertical lines are perpendicular. Find the angle labelled $x°$.

Solution
The angles along the line must add up to 180°. The two angles labelled $x°$ add up to 90°.
$2x = 90$, hence $x = 45$.
The angle is 45°.

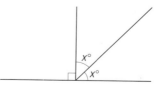

Fig. 13.4

EXERCISE 13A

1. Find the unknown angles in the diagrams of Fig. 13.5.

2. When a wheel goes round once, it turns through 360°. How many degrees correspond to:
   a) half a turn b) a quarter turn c) two turns?

3. When a wheel turns through the following angles, how many times has it turned?
   a) 720° b) 1800° c) 540°

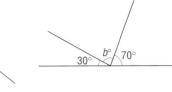

Fig. 13.5

4. A Ferris wheel turns a full circle every 5 minutes. How many degrees does it turn through in:
   a) one minute b) one second?

5. During two hours, what angle has the hour hand of a clock passed through?

6. The minute hand of a clock rotates through 90°. How much time has passed?

7. What is the angle between the hands of a clock at the following times:
   a) three o'clock b) six o'clock c) five o'clock?

8. Fig. 13.6 shows two straight lines crossing at a point. Find the unknown angles $x°$, $y°$ and $z°$.

9. In each of the following sentences, insert one of the words *acute*, *obtuse*, *right* or *reflex*.
   a) The angle 45° is ....
   b) Perpendicular lines meet at a ... angle.
   c) If $x°$ is acute, then $180° - x°$ is ....
   d) If $360° - y°$ is acute, then $y°$ is ....

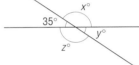

Fig. 13.6

Equal angles

Opposite angles — The angles on opposite sides of a crossing are equal.

Alternate angles — The angles on either side of a **transversal** (a line that crosses two parallel lines) are equal.

Corresponding angles — The angles on the same side of a transversal are equal.

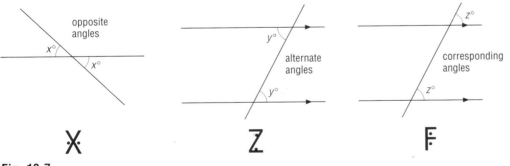

Fig. 13.7

Opposite, alternate and corresponding angles can be remembered as X angles, Z angles and F angles respectively, as shown in Fig. 13.7.

PLANE FIGURES

Example 13.2

Find the angles labelled $x°$ and $y°$ in Fig. 13.8

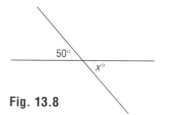
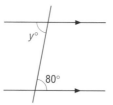

Fig. 13.8

Solution

The angle labelled $x°$ is opposite to the 50° angle.
$x° = 50°$
The angle labelled $y°$ is alternate to the angle of 80°.
$y° = 80°$

EXERCISE 13B

1. Find the angles labelled $a°$, $b°$ and $c°$ in Fig. 13.9.

2. In Fig. 13.10, the lines AB and CD are parallel. What is the angle labelled $x°$?

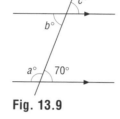

Fig. 13.9

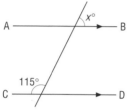

Fig. 13.10   **Fig. 13.11**

3. Fig. 13.11 shows two pairs of parallel lines. In each of the unmarked angles, write either $x°$ or $y°$. What is the relationship between x and y?

4. Fig. 13.12 is not drawn to scale. In which of the diagrams is there a pair of parallel lines?

a) 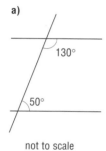 b) 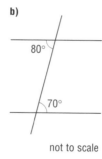 c)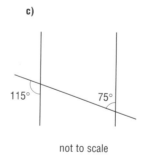

not to scale   not to scale   not to scale

Fig. 13.12

13.2 Polygons

A **polygon** is a plane (two-dimensional) figure with straight sides.

Names of polygons (Fig. 13.13)

Triangle	3 sides
Quadrilateral	4 sides
Pentagon	5 sides
Hexagon	6 sides

triangle

quadrilateral

pentagon

hexagon

Fig. 13.13

In a **regular polygon**, all the angles and sides are equal.

A **tessellation** is a pattern of polygons which covers a flat surface. Fig. 13.14 shows a tessellation of hexagons.

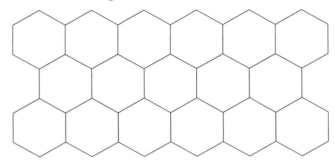

Fig. 13.14

Triangles (Fig. 13.15)

The angles of a triangle add up to 180°.

Isosceles   An isosceles triangle has two equal sides. The base angles are equal.

Equilateral   An equilateral triangle has all sides equal. Each angle is equal to 60°.

Right-angled   A right-angled triangle has one 90° angle.

Scalene   A scalene triangle has all sides different and all angles different.

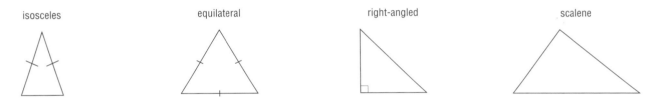

Fig. 13.15

Quadrilaterals (Fig. 13.16)

The angles of a quadrilateral add up to 360°.

Trapezium   A quadrilateral with one pair of parallel sides.

Parallelogram   A quadrilateral with two pairs of parallel sides. Opposite sides are equal and opposite angles are equal.

Rectangle   A quadrilateral with all angles equal to 90°. Opposite sides are equal and diagonals are equal.

Square   A quadrilateral with all sides equal and all angles equal to 90°. Diagonals are equal and perpendicular.

Rhombus   A quadrilateral with all sides equal. Opposite angles are equal and diagonals are perpendicular.

Kite   A quadrilateral with two pairs of adjacent sides equal and one pair of angles equal. The diagonals are perpendicular.

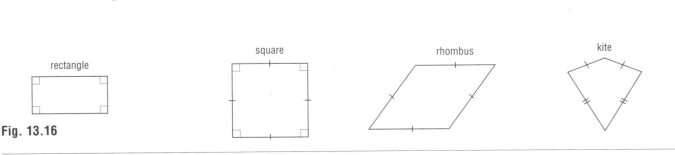

Fig. 13.16

Example 13.3

In Fig. 13.17, two sides of the triangle are equal. The top angle is 50°. What is the angle labelled $x°$?

Solution
The two base angles add up to 130°. The triangle is isosceles, so they must both be equal to $x°$. Divide 130 by 2, obtaining 65.
$x° = 65°$

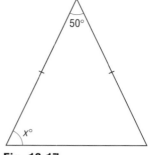

Fig. 13.17

Key point
■ Label a polygon correctly. The letters must go either clockwise or anti-clockwise around the figure. They should not jump across a diagonal.

EXERCISE 13C

1. Find the unknown angles in the triangles and quadrilaterals of Fig. 13.18.

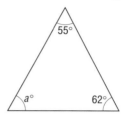 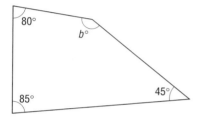 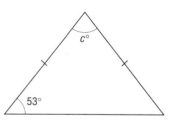 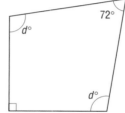

Fig. 13.18

2. In Fig. 13.19, AB = BC = CD and CÂB = 25°. Find the angles $x°$ and $y°$.

3. ABC is a triangle in which AB = AC and BĈA = 55°. Find BÂC.

4. LMN is a triangle in which LN̂M = LM̂N and NL̂M = 58°. Find the other angles of the triangle.

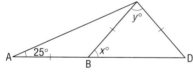

Fig. 13.19

5. Fig. 13.20 shows triangle XYZ, in which XY = XZ and XŶZ = 60°. What are the other angles of the triangle? What sort of triangle is △XYZ?

6. Extend the tessellations of Fig. 13.21, by drawing at least three more shapes on each.

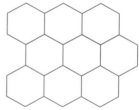 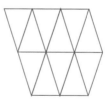

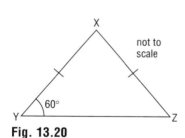

Fig. 13.20

Fig. 13.21

7. Fig. 13.22 shows the square ABCD. What sort of triangle is ABC?

8. Fig. 13.23 shows several quadrilaterals. Give each quadrilateral a name from the list below.

Fig. 13.22

kite   rhombus   rectangle   trapezium

a)   b)   c)   d)

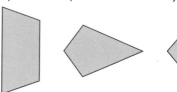 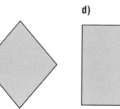 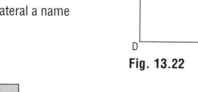

Fig. 13.23

Angles of a polygon

If a polygon has *n* sides (Fig. 13.24), the sum of its **interior angles** is $180° \times n - 360°$. This expression can also be written as $180°(n - 2)$.

The sum of its **exterior angles** is 360°.

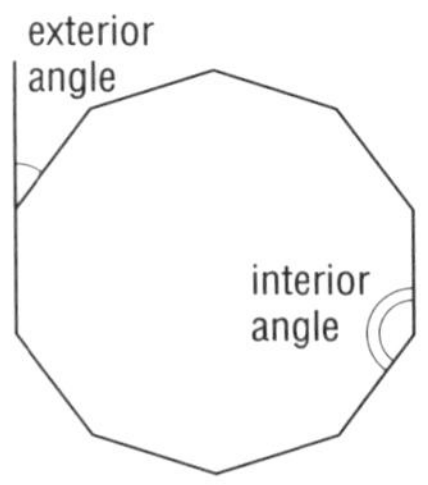

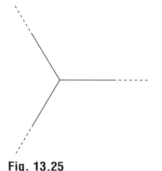

Fig. 13.24

■ Example 13.4

Find each interior angle of a regular octagon (8 sides).

Solution

Put $n = 8$ into the formula above, to find the sum of the interior angles.
$180° \times 8 - 360° = 1080°$
As the figure is regular, the interior angles are equal. Divide 1080 by 8, obtaining 135.
Each interior angle is 135°.

■ Example 13.5

A tessellation is made out of regular *n*-sided polygons. Three tiles meet at each vertex. Find *n*.

Solution

A vertex is shown in Fig. 13.25. The interior angle of each polygon is 360° divided by 3, i.e. 120°.
The exterior angle is therefore 60°. Divide 360 by 60, obtaining 6.
The polygon has six sides, i.e. is a hexagon.
$n = 6$.

Fig. 13.25

Key point

■ Don't assume that a polygon is regular, unless you are told it is.

EXERCISE 13D

1. Find the sum of the interior angles of:

   a) a hexagon   **b)** a heptagon (7 sides)   **c)** a decagon (10 sides)

2. Find each interior angle of a regular figure with:

   a) 5 sides   **b)** 20 sides   **c)** 9 sides

3. Find each exterior angle of a regular figure with:

   a) 6 sides   **b)** 12 sides   **c)** 18 sides

4. The sum of the interior angles of a polygon is 540°. What is the name for this figure?

5. Each exterior angle of a regular polygon is 60°. What is the name for this figure?

6. Each interior angle of a regular polygon is 160°. How many sides does it have?

7. Find the sum of the interior angles of a pentagon. These angles are x, $x - 10°$, $x + 10°$, $x + 20°$, $x + 30°$. Find x.

8. Not every angle can be the interior angle of a regular polygon. Which of the following can be the interior angle of a regular polygon? For each possible case give the number of sides of the polygon.

   a) 170°   **b)** 160°   **c)** 145°

9 Fig. 13.26 shows a regular pentagon ABCDE. Find the angles AB̂C, AĈB and AĈD. What sort of triangle is ACD?

10 Fig. 13.27 shows a regular hexagon ABCDEF. Find the angles AB̂C, AĈB and AĈD. What sort of quadrilateral is ACDF?

11 Leah's lucky number is 7. She wants to tile her kitchen floor with regular heptagons (7-sided figures). Explain to her why this is not possible. What regular polygons could she tile the floor with?

12 A floor is tiled with squares and with other regular polygons. Fig. 13.28 shows part of the tiling. Find the internal angle of the polygon, and hence find the number of its sides.

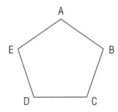
Fig. 13.26

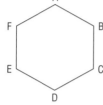
Fig. 13.27

Fig. 13.28

13.3 Circles

There are several special words connected with circles (Fig. 13.29).

Centre	The middle of the circle.
Radius	Line from a point on the circle to the centre.
Diameter	Distance across a circle through the centre. The diameter is twice the radius.
Circumference	The length around a circle.
Chord	Straight line joining two points on a circle.
Arc	Part of a circle between two points on the circle.
Sector	Region of a circle between two radii.
Segment	Region of a circle cut off by a chord.
Semi-circle	Half a circle.
Tangent	Straight line which just touches a circle.

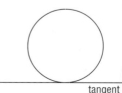

Fig. 13.29

PLANE FIGURES

Circle facts
- All radii are equal in length.
- The angle between a radius and a tangent is 90°.
- If AB is a diameter, and C a point on the circle, then $A\hat{C}B = 90°$.

■ Example 13.6
A and B are points on a circle with centre C. $A\hat{C}B = 70°$. Find the other angles of the triangle ABC.

Solution
See Fig. 13.31. CA and CB are radii of the same circle. Hence they are equal.
The angles $C\hat{A}B$ and $C\hat{B}A$ are equal because the triangle is isosceles.
Subtract 70 from 180, obtaining 110. Divide by 2, obtaining 55.
The other angles are both 55°.

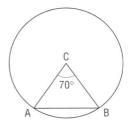
Fig. 13.31

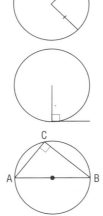
Fig. 13.30

EXERCISE 13E

1. In Fig. 13.32 there are six circles. Use them to indicate:
   a) the centre
   b) a radius
   c) a diameter
   d) a sector
   e) a segment
   f) a semicircle

2. On the circles of Fig. 13.33 indicate:
   a) the chord XY
   b) the arc XY

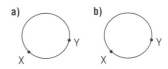
Fig. 13.33

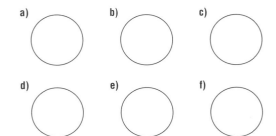
Fig. 13.32

3. In Fig. 13.34, A, B and C are points on a circle, and AB is a diameter. If $C\hat{A}B = 35°$ find $A\hat{C}B$ and $C\hat{B}A$.

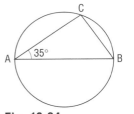
Fig. 13.34

4. In Fig. 13.35, AB is a diameter of a circle, and AX is a tangent. If $A\hat{X}B = 53°$ find $A\hat{B}X$.

5. Find the unknown angles $a°$ and $b°$ in the circles of Fig. 13.36. The centres of the circles are labelled X.

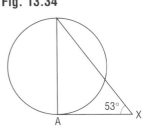
Fig. 13.35

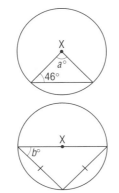
Fig. 13.36

Revision checklist
This chapter has revised:

13.1 Angles and lines. Names for angles, pairs of equal angles. ❑
13.2 Polygons. Special triangles, special quadrilaterals. The sum of the angles of a polygon. ❑
13.3 Circles. Names for points, lines and regions connected with a circle. Facts about equal radii and angles equal to 90°. ❑

© IT IS ILLEGAL TO PHOTOCOPY THIS PAGE

CHAPTER 14 *Measures, lengths and areas*

> **Chapter key points**
> - Be careful when converting between units, especially between the metric system and the Imperial system.
>   - To go from pounds to grams, *multiply* by 454.
>   - To go from grams to pounds, *divide* by 454.
> - Be careful with distance, speed and time. Make sure that you divide and multiply in the right situations, as follows:
>
> $$\text{speed} = \text{distance} \div \text{time}$$
> $$\text{distance} = \text{speed} \times \text{time}$$
> $$\text{time} = \text{distance} \div \text{speed}$$
>
> - Read questions on circles carefully to see whether the radius or diameter of a circle is referred to.
> - Given the diameter, get the circumference by *multiplying* by π. Given the circumference, get the diameter by *dividing* by π.
> - In the formula $A = \pi r^2$ for the area of a circle, square first and then multiply by π, not the other way round.

14.1 Units and change of units

The two systems of units in common use are the **metric** system and the **Imperial** system. The most important units in the systems are given below. Abbreviations are in brackets. The approximate conversion rates are also given.

Length

Metric
The basic unit is the **metre** (m).
1 metre = 100 centimetres (cm)
1 metre = 1000 millimetres (mm)
1000 m = 1 kilometre (km)

Imperial
12 inches (in) = 1 foot (ft)
3 feet = 1 yard (yd)
1760 yds = 1 mile

- Conversion:

    1 inch ≃ 2.54 cm

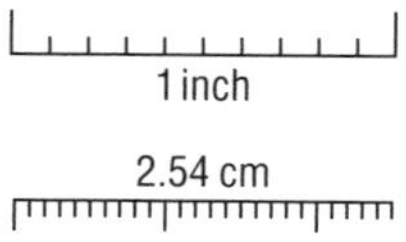

Fig. 14.1

Weight

Metric
The basic unit is the **gram** (g)
1 gram = 1000 milligrams (mg)
1000 grams = 1 kilogram (kg)

Imperial
16 ounces (oz) = 1 pound (lb)
14 lb = 1 stone
160 stones = 1 ton

- Conversion

    1 pound ≃ 454 grams

Area

Metric
10 000 cm^2 = 1 m^2
10 000 m^2 = 1 hectare (ha)

Imperial
4840 square yards = 1 acre

MEASURES, LENGTHS AND AREAS

■ Conversion:

1 acre ≃ 4047 square metres

Volume

Metric
The basic unit is the **litre** (l).
1 litre = 100 centilitres (cl)
1 litre = 1000 millilitres (ml)
1 millilitre = 1 cubic centimetre
  (c.c. or cm^3)

Imperial
2 pints = 1 quart
4 quarts = 1 gallon

■ Conversion:

1 gallon ≃ 4.54 litres

■ Example 14.1

An American cousin tells you that she weighs 115 lb.
a) What is her weight in stones and pounds?
b) What is her weight in kg?

Solution
a) There are 14 pounds in 1 stone. Divide 115 by 14, obtaining 8 with a remainder of 3.
She weighs 8 stones 3 pounds.
b) One pound is 454 grams, i.e. 0.454 kg. Multiply 0.454 by 115, obtaining 52.21.
She weighs 52.21 kg.

■ Example 14.2

The width of a building is 15 m. What is its width in feet? Give your answer correct to 3 significant figures.

Solution
The width is 1500 cm. To convert to inches, *divide* by 2.54. Then divide by 12, to convert to feet.
1500 ÷ 2.54 = 590.55
590.55 ÷ 12 = 49.2
The width is 49.2 ft.

Key point

■ Be careful when converting between units, especially between the metric system and the Imperial system.

- To go from pounds to grams, *multiply* by 454.
- To go from grams to pounds, *divide* by 454.

EXERCISE 14A

1 Make the following conversions within the metric system.

a) 5 metres to centimetres
b) 1250 g to kg
c) 30 000 m^2 to hectares
d) 0.223 l to ml

2 Make the following conversions within the Imperial system.
   a) 53 yards to feet
   b) 70 pounds to stones
   c) 1.5 miles to yards
   d) 100 pints to gallons

3 Make the following conversions between the metric and Imperial systems.
   a) 25 inches to cm
   b) 12 gallons to litres
   c) 24 kg to pounds
   d) 28 hectares to acres

4 The chart of Fig. 14.2 converts between miles and kilometres. Use the chart for the following:
   a) the distance in km of 14 miles
   b) the distance in miles of 26 km

5 The floppy disks used by a computer are $3\frac{1}{2}$ in wide. What is their width in mm?

6 The length of a room is 18 feet. Convert this:
   a) to inches
   b) to metres

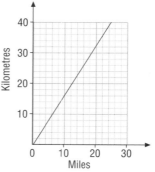

Fig. 14.2

Mixed measures

Some measurements involve more than one quantity.

- **Density** $= \dfrac{\text{mass}}{\text{volume}}$ e.g. grams per cubic centimetre (g/cm^3)

- **Speed** $= \dfrac{\text{distance}}{\text{time}}$ e.g. miles per hour (m.p.h.) or metres per second (m/s)

■ Example 14.3

I drive for 100 miles at 40 m.p.h. Then I drive for $1\frac{1}{2}$ hours at 70 m.p.h. Find:
a) the total distance travelled
b) the total time taken
c) the average speed of the whole journey.

Solution

a) Multiply time and speed for the distance of the second half of the journey.
   $1\frac{1}{2} \times 70 = 105$
   Add this to 100.
   The total distance was 205 miles.

b) Divide distance by speed for the time of the first half of the journey.
   $100 \div 40 = 2\frac{1}{2}$
   Add this to $1\frac{1}{2}$.
   The total time was 4 hours.

c) For the average speed, divide the total distance by the total time.
   $205 \div 4 = 51.25$
   The average speed was 51.25 m.p.h.

Key point

■ Be careful with distance, speed and time. Make sure that you divide and multiply in the right situations, as follows:

   speed = distance ÷ time
   distance = speed × time
   time = distance ÷ speed

If in doubt, think of an actual journey that you are familiar with. The triangle of Fig. 14.3 may help.

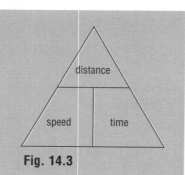

Fig. 14.3

MEASURES, LENGTHS AND AREAS

EXERCISE 14B

1. A sea-crossing of 150 miles takes $7\frac{1}{2}$ hours. What is the average speed?

2. Sidney drives for two hours at 50 m.p.h., then he drives 30 miles at 60 m.p.h.
   a) What is his total distance?
   b) What is his total time?
   c) What is his average speed?

3. A woman walks at a steady speed of 6 km per hour.
   a) How far does she walk in 4 hours?
   b) How long does she take to cover 15 km?

4. The mass of 2.5 m³ of a liquid is 2700 kg. What is its density, in kg/m³?

5. The density of a metal is 9 g/cm³.
   a) What is the mass of 50 cm³ of the metal?
   b) What is the volume of 54 g of the metal?

14.2 Length

The **perimeter** of a shape is the length round it.

- For a rectangle, length l and breadth b, perimeter $= 2l + 2b$
- For a circle, diameter d or radius r, perimeter (circumference) $= \pi d$ or $2\pi r$

Here π is a number which is approximately 3.14 or $\frac{22}{7}$. A calculator has a key which gives π more accurately.

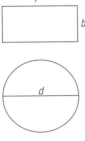

Fig. 14.4

■ Example 14.4
The radius of a compact disc is 6 cm. What is its perimeter?

Solution
Use the formula, perimeter $= 2\pi r$.
Perimeter $= 2 \times \pi \times 6 = 37.7$ cm

Key points
- Read questions on circles carefully, so see whether the radius or the diameter of a circle is referred to.
- Given the diameter, get the circumference by *multiplying* by π.
  Given the circumference, get the diameter by *dividing* by π.

EXERCISE 14C

1. A square has side 30 cm. What is its perimeter?

2. The radius of a circular running track is 60 m.
   a) What is the diameter of the track?
   b) How long is one circuit of the track?

3. The diameter of a floppy disk is $3\frac{1}{2}$ inches. What is the perimeter of the disk?

4. The minute hand of a watch is 1.8 cm long. In one hour, how far does the tip of the hand travel?

5. Fig. 14.5 shows the plan of a room. The walls meet at right angles. What is the perimeter of the room?

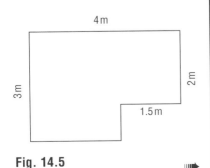

Fig. 14.5

6   A circular leisure pool is to have perimeter 500 m. What will the diameter of the pool be?
7   The perimeter of the circular top of a tin is 45 cm. Find its radius of the top.
8   The dots of Fig. 14.6 are 1 cm apart. Find the perimeter of the shape.

Fig. 14.6

14.3 Area

The areas of basic shapes are given below (Fig. 14.7).

Square, Area = x^2 Rectangle, Area = lb Parallelogram, Area = bh Triangle, Area = $\frac{1}{2}bh$ Trapezium, Area = $\frac{1}{2}h(a + b)$ Circle, Area = πr^2

    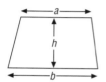

Fig. 14.7

■ Example 14.5

In Fig. 14.8 the dots are 1 cm apart. Find the area of the parallelogram.

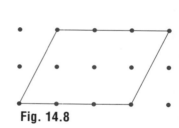

Fig. 14.8

Solution
The base of the parallelogram is 3 cm. Its height is 2 cm. Use the formula given above.
$3 \times 2 = 6$
The area is 6 cm².

■ Example 14.6

The radius of a compact disc is 6 cm. Find the area of its top surface.

Solution
Use the formula for the area of a circle.
$A = \pi r^2 = \pi \times 6^2 = 113$
The area is 113 cm².

Key point
■ In the formula $A = \pi r^2$ for the area of a circle, square first and then multiply by π, not the other way round.

EXERCISE 14D

1   The dots of Fig. 14.9 are 1 cm apart. Find the areas of the shapes shown.

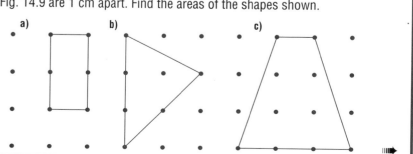

Fig. 14.9

2. The floor of a room is a rectangle which is 3 m by $2\frac{1}{2}$ m. It will be carpeted, using carpet at £10.50 per square metre. What will be the cost of the carpet?

3. A field is a rectangle 40 m by 30 m.
   a) What is the area of the field?
   b) What is the perimeter of the field?
   c) Fertiliser at 12p per square metre is to be spread on the field. How much will this cost?
   d) A fence costing £13 per metre is to be put around the field. What will the fence cost?

4. A rectangle has area 4000 m², and one side is 50 m. What is the length of the other side?

5. The base of a triangle is 5 cm, and its area is 25 cm². What is the height of the triangle?

6. Fig. 14.10 shows one side of a swimming pool. The diagram is not to scale. What is the area of the side?

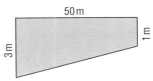

Fig. 14.10

7. Fig. 14.11 shows the side of a garden shed, which is 2 m from back to front. The back wall is $1\frac{1}{2}$ m high and the roof slopes uniformly to the front wall which is $2\frac{1}{2}$ m high. Find the area of the side wall.

8. Find the area of the circles with:
   a) radius 12 m
   b) diameter 14 cm

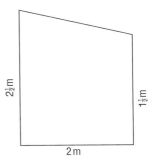

Fig. 14.11

14.4 Combinations of shapes

■ Example 14.7
Find the area of the L-shape of Fig. 14.12. Lengths are in cm.

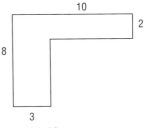

Fig. 14.12

Solution
Split the shape into two rectangles, one 8 by 3, the other 7 by 2.
$8 \times 3 + 7 \times 2 = 38$
The area is 38 cm².

■ Example 14.8
A window consists of a semi-circle on top of a square of side 42 cm. (Fig. 14.13). Find the area of the window.

Solution
The semi-circle has radius 21 cm.
Area of square = 42² cm² = 1764 cm²
Area of semi-circle = $\frac{1}{2} \times \pi \times 21^2$ cm² = 693 cm²
Total area = 2457 cm²

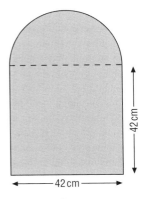

Fig. 14.13

EXERCISE 14E

1. Find the areas of the shapes in Fig. 14.14. Lengths are in m.

   a) 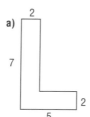   b) 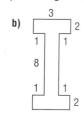   c)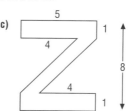

   Fig. 14.14

2. In Fig. 14.15 the dots are 1 cm apart. Find the areas of the shapes.

   a) 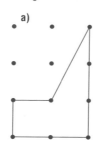   b) 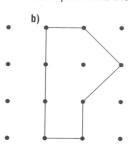   c)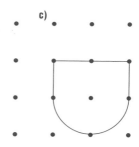

   Fig. 14.15

3. Fig. 14.16 shows a running track. The straight sections are 60 m long, and the curved sections are semi-circles of radius 30 m.

   a) Find the total perimeter.
   b) Find the total area enclosed by the track.

   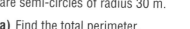

   Fig. 14.16

4. Find the area of the L-shaped room shown in Fig. 14.5.

5. Fig. 14.17 shows the side wall of a swimming pool. The diagram is not to scale. Find the area of the side wall.

   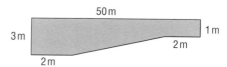

   Fig. 14.17

Revision checklist

This chapter has revised:

14.1 The metric and Imperial systems of measures. Conversion within each system, and conversion between the systems. Mixed measures such as speed and density. ☐

14.2 Perimeters of basic shapes. ☐

14.3 Areas of basic shapes. ☐

14.4 Areas of combinations of shapes. ☐

CHAPTER 15 Solids and volumes

Chapter key points
- Try to make your diagrams of solids reasonably clear so that they can be understood. For example, it should be clear which lines are parallel to each other. But don't spend too long on this – you are not being tested for your drawing skills.
- Numbers like 3 or π have no dimension.
- Suppose a and b are lengths. Then $a \times b$ is an area. But $a + b$ is another length – think of two sticks laid end to end.

15.1 Types of solids

There are some special words used to describe solids (three-dimensional objects) (Fig. 15.1).

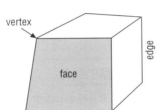

Fig. 15.1

Face	A flat side of a solid.
Edge	A line on a solid where two faces meet.
Vertex	A point on a solid where three or more edges meet.
Cross-section	A slice through a solid, usually parallel to a face.

The names and properties of basic solids are given below (Fig 15.2).

Cube	6 square faces
Cuboid	6 rectangular faces
Prism	constant cross-section
Triangular prism	prism with triangular cross-section
Pyramid	tapers to a point from a rectangular base
Tetrahedron	4 triangular faces
Sphere	round solid like a ball
Cylinder	cross-section is a circle
Cone	tapers to a point from a round base

cube

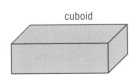
cuboid

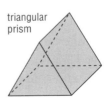
triangular prism

pyramid

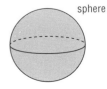
tetrahedron

sphere

cylinder

cone

Fig. 15.2

SOLIDS AND VOLUMES

■ Example 15.1

Fig. 15.3 shows a triangular prism.

a) Write down the number of its faces, edges and vertices.
b) Name a pair of parallel faces.
c) Name three edges that are parallel to each other.

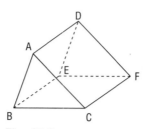

Fig. 15.3

Solution

a) There are two triangular faces and three rectangular faces.
   There are three edges on each triangle and three more edges joining the triangles.
   There are three vertices at each end.
   There are 5 faces, 9 edges and 6 vertices.
b) The triangles at each end are parallel.
   ABC and DEF are parallel.
c) The three lines joining the triangles are parallel.
   AD, BE and CF are parallel.

EXERCISE 15A

1. Find the number of faces, edges and vertices of a pyramid (Fig. 15.4).

2. The cube ABCDEFGH is shown in Fig. 15.5.
   a) Write down the number of its faces, edges and vertices.
   b) Name a pair of parallel faces.
   c) Name a pair of parallel edges.

3. What are the mathematical names for the following common objects?
   a) a shoe polish tin
   b) a plank
   c) a tennis ball
   d) a wedge used to stop a door

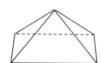

Fig. 15.4

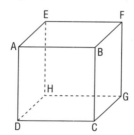

Fig. 15.5

4. Fig. 15.6 shows a cube of side 3 cm, which is made from 27 sugar cubes of side 1 cm. Each sugar cube has either 0, 1, 2 or 3 faces exposed to the air. How many have:
   a) 0 exposed faces
   b) 1 exposed face
   c) 2 exposed faces
   d) 3 exposed faces?

5. Suppose a solid with flat faces has f faces, e edges and v vertices. A theorem states that $f - e + v$ always has the same value.
   a) Use Example 15.1 to find the value.
   b) Check the value on some other solids described above.

Fig. 15.6

15.2 Drawing and making solids

Isometric paper is covered by equilateral triangles. Cuboids can be drawn on this sort of paper, as shown in Example 15.2 below.

A **net** is a diagram that can be cut out and folded to make a solid, as shown in Example 15.3.

■ Example 15.2

A cuboid is 2 units by 3 units by 2 units. Draw the cuboid on isometric paper.

Solution

Draw a vertical line of length 2 units. From both ends draw sloping lines of 3 units and of 2 units. Complete the cuboid as shown in Fig. 15.7.

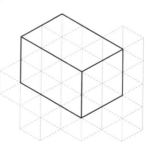

Fig. 15.7

Example 15.3

Fig. 15.8 shows the net of a solid.

a) What is the name of the solid?
b) Which points will F join?

Solution

a) When the solid is assembled, the four triangles ABE, BCF, CDG and DAH will meet at a point.
   The solid is a pyramid.
b) F will join the tips of the other triangles.
   F will join E, G and H.

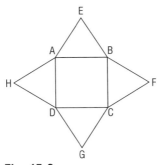

Fig. 15.8

Key point

■ Try to make your diagram of solids reasonably clear, so that they can be understood. For example, it should be clear which lines are parallel to each other. But don't spend too long on this – you are not being tested for your drawing skills.

EXERCISE 15B

1. On isometric paper, draw a cuboid which is 2 units by 4 units by 3 units.

2. On isometric paper, draw a pyramid, 2 units high, with a rectangular base which is 4 units by 3 units.

3. A prism is 5 units long. Its cross-section is a triangle, with sides of length 3 units and 4 units at right-angles to each other (Fig. 15.9). Draw this prism on isometric paper.

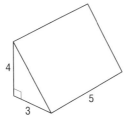

Fig. 15.9

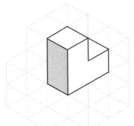

Fig. 15.10

4. Fig. 15.10 shows a shape on isometric paper. It is tipped over so that the shaded side lies on the paper. Draw the shape as it now appears.

5. The dice used in games are always marked so that the numbers of dots on opposite faces add up to 7. Fig. 15.11 shows the net of a die. Fill in the blank faces.

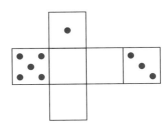

Fig. 15.11

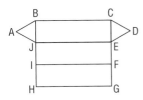

Fig. 15.12

6. Fig. 15.12 shows the net of a solid.

   a) What solid will be formed from the net?
   b) When the solid is formed, what point will join A?

7 Darren wants to make a cube out of a net. He draws several patterns of squares, as shown in Fig. 15.13. Which of them will work?

a)   b)   c)   d)   e)

Fig. 15.13

8 Draw accurate nets for the solids in Fig 15.14.
  a) a cuboid 3 cm by 4 cm by 5 cm
  b) a pyramid, square base side 4 cm, slanting edges of length 5 cm

a) 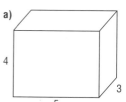   b)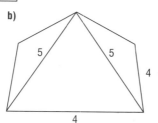

Fig. 15.14

15.3 Volume

The volumes of basic solids are given below (Fig 15.15).

Cube, side x: x^3

Cuboid, a by b by c: abc

Prism, cross-section area A, height h: Ah

Pyramid, base area A, height h: $\frac{1}{3}Ah$

Cylinder, base radius r, height h: $\pi r^2 h$

 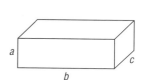 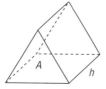

Fig. 15.15

■ Example 15.4

A can is 11 cm high and has base radius 4 cm. Find the volume of the can.

Solution
The can is a cylinder, with $h = 11$ and $r = 4$. Use the formula given above.
$\pi \times 4^2 \times 11 = 553$
The volume is 553 cm^3.

■ Example 15.5

An enclosed concrete yard is a rectangle, 30 m by 40 m. During a storm, 3 cm of water falls. Find the volume of water.

Fig. 15.16

Solution
The water can be thought of as a cuboid, 30 m by 40 m by 3 cm.
Convert 3 cm to 0.03 m and multiply the numbers together.
$30 \times 40 \times 0.03 = 36$
The volume of water is 36 m^3.

EXERCISE 15C

1. A cuboid is 3 cm by 4 cm by 4.5 cm. Find its volume.

2. Squash balls are spheres with diameter 4 cm. They are packed in cubes made from thin cardboard. What is the volume of these cubes?

3. A pond is a rectangle 5 m by 6 m. In winter it freezes to a depth of 1 cm. What is the volume of the ice?

4. A room is a cuboid, of volume 126 m^3. Its floor is a rectangle 7 m by 6 m. What is the height of the room?

5. The dots of Fig. 15.17 are 1 cm apart. What solids are formed from the nets shown? What are their volumes?

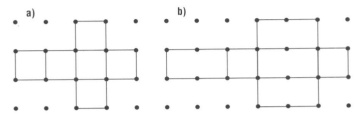

Fig. 15.17

6. A can of soup is 17 cm high, and its base has diameter 8 cm. Find the volume of the can.

7. A room is 280 cm high, 500 cm long and 420 cm wide.
   a) What is the total area of the walls and ceiling?
   b) A coat of paint is 0.01 cm thick. How much paint is needed to give one coat on the walls and ceiling?

8. The side wall of a shed is a trapezium. The parallel vertical sides are 2 m and $2\frac{1}{2}$ m. The distance from the back of the shed to the front is $1\frac{3}{4}$ m. See Fig. 15.18.
   a) What is the area of a side wall?
   b) If the shed is $3\frac{1}{4}$ m long, find its volume.

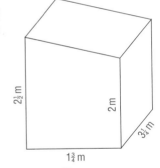

Fig. 15.18

15.4 Combinations of solids

■ Example 15.6

A hut consists of half a cylinder on top of a cuboid. The cylinder has diameter 4 m and length 5 m. The cuboid is 4 m by 5 m by 3 m. The hut is shown in Fig. 15.19. Find its volume.

Solution

Find the area of an end wall. This is a semi-circle on top of a rectangle.
Area of semi-circle $= \frac{1}{2} \times \pi \times 2^2 = 6.28$ m^2
Area of rectangle $= 3 \times 4 = 12$ m^2
Surface area of end wall $= 18.28$ m^2
Multiply the area by the length, 5 m, to obtain the volume.
The volume is 91.4 m^3.

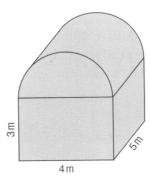

Fig. 15.19

EXERCISE 15D

1. Fig. 15.20 shows the side wall of a swimming pool. The diagram is not to scale.
   a) Find the area of the wall.
   b) If the pool is 10 m wide, find its volume.

2. Fig. 15.21 shows the cross-section of a steel girder.
   a) Find the area of cross-section.
   b) If the girder is 400 cm long, find its volume.

3. Fig. 15.22 shows the end wall of a hut, which is a triangle on top of a rectangle.
   a) Find the area of the wall.
   b) If the hut is 1.8 m long, find its volume.

4. The cross-section of a copper pipe is shown in Fig. 15.23. The inside radius is 1.8 cm and the outside radius is 2 cm.
   a) By subtracting the area of one circle from another, find the shaded area.
   b) Find the volume of copper in 3000 cm of the pipe.

5. Fig. 15.24 shows a concrete pillar. It consists of a pyramid on top of a cuboid. The cuboid has height 120 cm and a square base of side 40 cm. The total height is 150 cm. Find the volume of the pillar.

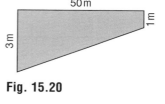

Fig. 15.20

Fig. 15.21

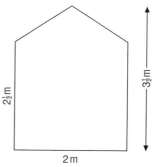

Fig. 15.22

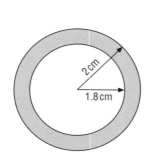

Fig. 15.23

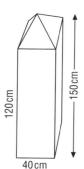

Fig. 15.24

15.5 Dimensions

Length, area and volume have 1 dimension, 2 dimensions and 3 dimensions, respectively. You can look at a formula and tell whether it refers to length, area or volume. The product of two lengths, for example, must be an area.

■ Example 15.7

Suppose a cylinder has radius r and height h. Does the formula $2\pi r(r + h)$ represent length, area or volume?

Solution

2 and π are numbers without dimension; r and $(r + h)$ are both lengths, hence their product is an area.
The formula represents area.

Key points

■ Numbers like 3 or π have no dimension.
■ Suppose a and b are lengths. Then $a \times b$ is an area. But $a + b$ is another length – think of two sticks laid end to end.

EXERCISE 15E

1 In the formulae below, x and y are lengths, A is an area. Do the formulae represent length, area or volume?

a) xy **b)** xA **c)** $2x + 3y$

d) $2y^2x + x^3$ **e)** $4\pi x^2$ **f)** $\dfrac{A}{x+y}$

2 Fig. 15.25 shows a cuboid. Complete the following sentences with the words *perimeter*, *area* or *volume*.

a) $2(ab + bc + ca)$ is the ... of the cuboid.
b) $2(a + b)$ is the ... of the base.
c) abc is the ... of the cuboid.

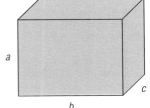

Fig. 15.25

3 Fig. 15.26 shows a cone of height h and base radius r. The following formulae are connected with the cone. Do they represent length, area or volume?

a) $\tfrac{1}{3}\pi r^2 h$
b) $\sqrt{(r^2 + h^2)}$
c) $\pi r \sqrt{(r^2 + h^2)}$

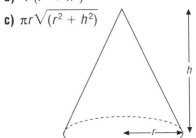

Fig. 15.26

Revision checklist

This chapter has revised:

15.1 Types of solid, their faces, edges and vertices. ❑
15.2 Drawing solids on isometric paper. Making nets for solids. ❑
15.3 The volumes of basic solids. ❑
15.4 The volumes of combinations of solids. ❑
15.5 Determining the dimensions of a formula (whether it represents length, area or volume). ❑

CHAPTER 16 Construction and maps

Chapter key points

- In Maths, you are not tested for your skill in drawing. But do the best you can – use a sharp pencil, a straight ruler and a non-wobbly compass!
- The faint lines you make during a construction are part of your answer, and the examiner wants to see them. Don't rub them out.
- The examiner is expecting reasonable accuracy in your answers. Measure distances to the nearest mm, and angles to the nearest degree.
- Read a question carefully, to see whether you are asked the bearing of A from B or of B from A. To get the bearing of A from B, imagine yourself standing at B and looking towards A.
- Bearings are always measured from North. Even if the direction is closer to South, you still measure the bearing from North.
- Bearings are measured *clockwise* round from North. For example, due West is on a bearing of 270°.
- Read questions on locus carefully.
  - If X is a constant distance from a point, X lies on a circle.
  - If X is a constant distance from a line, X lies on another line.

16.1 Measurement and construction

The instruments used in geometrical drawings are:

Ruler to measure lengths (to an accuracy of about 1 mm) and draw straight lines
Protractor to measure and construct angles (to an accuracy of about 1°)
Compasses to draw circles

■ Example 16.1

A distance on a map is measured with a ruler, as shown in Fig. 16.1. What is the distance between X and Y?

Solution
The point X is at 3.8 cm. The point Y is at 8.5 cm. Subtract one measurement from the other, obtaining 4.7 cm.
The distance is 4.7 cm.

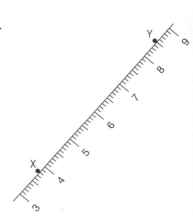

Fig. 16.1

■ Example 16.2

Fig. 16.2 shows three points A, B and C. Measure the angle $A\hat{B}C$. Point D is such that $B\hat{A}D = 57°$, and $B\hat{C}D = 63°$. Find the position of D.

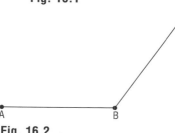

Fig. 16.2

Solution
Use a protractor to measure the angle at B.
$A\hat{B}C = 125°$
Use the protractor to construct a line making 57° with BA. Then construct a line making 63° with CB. These lines cross at D, as shown in Fig. 16.3.

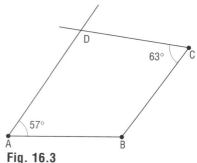

Fig. 16.3

CONSTRUCTION AND MAPS

Key points

- In Maths, you are not tested for your skill in drawing. But do the best you can – use a sharp pencil, a straight ruler and a non-wobbly compass.
- The faint lines you make during a construction are part of your answer, and the examiner wants to see them. Don't rub them out.
- The examiner is expecting reasonable accuracy in your answers. Measure distances to the nearest mm, and angles to the nearest degree.

EXERCISE 16A

1. In Fig. 16.4 a sphere is held between two blocks on top of a ruler. The ruler's scale is in cm. What is the diameter of the sphere?

2. Measure the angles $A°$, $B°$ and $C°$ in Fig. 16.5. Check that $A + B + C = 180$.

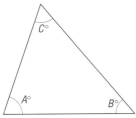

Fig. 16.5

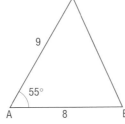

Fig. 16.6

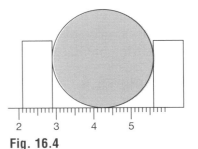

Fig. 16.4

3. Fig. 16.6 (not to scale) shows a triangle ABC, with AB = 8 cm, AC = 9 cm and $C\hat{A}B = 55°$.
   Construct this triangle accurately.
   Find the length BC and the angle $A\hat{C}B$.

4. Construct a triangle ABC for which AB = 7 cm, AC = 6.4 cm and $B\hat{A}C = 65°$. Measure the length BC and the angle $A\hat{B}C$.

5. The dots of Fig. 16.7 are 1 cm apart.
   a) Measure the length AC.
   b) Measure the angle $A\hat{C}B$.

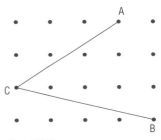

Fig. 16.7

6. Fig. 16.8 (not to scale) shows a triangle XYZ for which XY = 6 cm, $Z\hat{X}Y = 73°$, $Z\hat{Y}X = 42°$. Construct this triangle accurately.
   Measure the lengths XZ and YZ.

7. Construct a triangle LMN for which MN = 7.8 cm, $L\hat{M}N = 68°$ and $L\hat{N}M = 52°$.
   Measure the length LM.

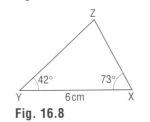

Fig. 16.8

16.2 Maps

The **scale** of a map gives the ratio of distance on the map to distance on the ground. A scale of 1 cm to 2 km means that 1 cm on the map represents 2 km on the ground:

    1 cm : 2 km   or   1 : 200 000

A **bearing** gives the direction of one point from another. Bearings are measured from North, in a clockwise direction. See Fig. 16.9. They are always given in three figures, e.g. 20° from North is a bearing of 020°.

If you are given the bearing of A from B, then the bearing of B from A is found by adding or subtracting 180°.

Fig. 16.9

© IT IS ILLEGAL TO PHOTOCOPY THIS PAGE

Example 16.3

Fig. 16.10 shows a map of part of Scotland. The scale is 1 cm : 10 km.

a) How far is it from Glasgow to Alloa?
b) What is the bearing of Alloa from Glasgow?
c) What is the bearing of Glasgow from Alloa?

Solution

a) Measure the distance on the map. It is 4 cm. Multiply this by 10.
   The distance from Glasgow to Alloa is 40 km.
b) The line from Glasgow to Alloa makes 45° with the North direction.
   The bearing of Alloa from Glasgow is 045°.
c) There is no need to measure an angle. Add 180 to 45, obtaining 225.
   The bearing of Glasgow from Alloa is 225°.

Fig. 16.10

Example 16.4

Some scouts are on an orienteering exercise. From their base, they travel South for 40 km, then for 30 km on a bearing of 058°. Use a scale of 1 cm to 10 km to make an accurate drawing of the journey. How far are they from their base?

Solution

Fig. 16.11 shows the journey. The first stage of the journey is the 4 cm line AB. The second stage is the 3 cm line BC, which makes 58° with the North–South line.
Measure AC, as 3.5 cm. Multiply by 10.
They are 35 km from their base.

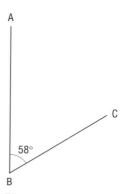

Fig. 16.11

Key points

- Read a question carefully, to see whether you are asked the bearing of A from B or of B from A. To get the bearing of A from B, imagine yourself standing at B and looking towards A.
- Bearings are always measured from North. Even if the direction is closer to South, you still measure the bearing from North.
- Bearings are measured *clockwise* round from the North. For example, due West is on a bearing of 270°.

EXERCISE 16B

1. Fig. 16.12 is part of a map showing two villages in Sussex. The scale is 1 cm per 2 km.

   a) Find the distance from Ticehurst to Wadhurst.
   b) What is the bearing of Ticehurst from Wadhurst?
   c) What is the bearing of Wadhurst from Ticehurst?

2. Fig. 16.13 shows part of a map of Malawi. The scale is 1 cm : 50 km.

   a) What is the distance of Lilongwe from Zomba?
   b) What is the bearing of Lilongwe from Zomba?

3. A plane leaves an airfield and flies North for 200 km. It then turns West and flies for another 150 km. Fig. 16.14 (not to scale) shows the journey of the plane.

   a) Make an accurate map of the plane's journey, using a scale of 1 cm per 20 km.
   b) How far is the plane from the airfield?
   c) What is the bearing of the plane from the airfield?

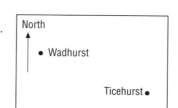

Fig. 16.12   Fig. 16.13

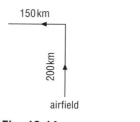

Fig. 16.14

4 Alison and Ben want to find the distance of a church. Ben stands 50 m East of Alison. The bearings of the church from Alison and Ben are 020° and 330° respectively. Fig. 16.15 (not to scale) shows the situation.

a) Draw an accurate map of the triangle ABC, using a scale of 1 cm to 10 m.
b) How far is Alison from the church?

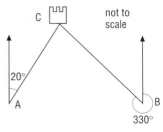

Fig. 16.15

5 Fig. 16.16 (not to scale) shows a journey made by Cyril. Originally he sees a church spire on a bearing of 040°. He walks North for 400 m, and then the spire is on a bearing of 170°.

a) Make an accurate map of Cyril's journey, using a scale of 1 cm per 20 m.
b) At the end of his journey, how far is Cyril from the spire?

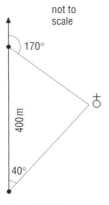

Fig. 16.16

16.3 Locus

The **locus** of a point is the path which it traces. Types of locus are listed below.

- Constant distance from line — If P is always 1 cm above a straight line, its locus is another straight line, 1 cm away.

- Constant distance from point — If P is always 1 cm from a point A, its locus is a circle with centre A and radius 1 cm.

- Equal distance between points — If P moves so that PA = PB, its locus is perpendicular to AB and cuts it in half.

- Equal distance between lines — If P moves so that it is the same distance from AB as from AC, then its locus cuts the angle CÂB in half.

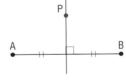

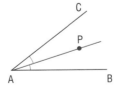

Fig. 16.17

■ Example 16.5

A field is a square ABCD of side 30 m. A farmer starts from A, walking so that he is an equal distance from AB and AD, and stops when he is within 5 m of C. Draw a diagram, scale 1 cm to 10 m, showing his journey.

Solution

First draw a square of side 3 cm. The farmer's locus cuts the angle BÂD in half, hence it makes 45° with AB. Draw this line.
The points within 5 m of C are represented by a circle, centre C, with radius 0.5 cm. Draw this circle. The farmer stops when the line cuts the circle, as shown on the diagram of Fig. 16.18.

Fig. 16.18

Example 16.6

Fig. 16.19 shows two countries, A and B, separated by sea. There is a treaty allowing ships from country A to fish as long as they are closer to point P than to point Q. Indicate on the map the region within which ships from country A can fish.

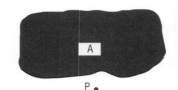

Solution
The border of the region is the line that cuts PQ in half, at right angles to PQ. Put the point of your compasses on Q, and make arcs on either side of PQ. Do the same with P, so that the arcs cross.
Join the crossing points. The region is above the dotted line of Fig. 16.20.

Fig. 16.19

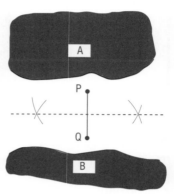

Key point
- Read questions on locus carefully.
  - If X is a constant distance from a point, X lies on a circle.
  - If X is a constant distance from a line, X lies on another line.

Fig. 16.20

EXERCISE 16C

1. Draw a square ABCD with side 10 cm.
   a) Find the points which are 4 cm from A.
   b) Find the points which are 2 cm from BC.
   c) Find the point which is 6 cm from A and 7 cm from B.
   d) Find the point which is 3 cm from AB and 4 cm from BC.

2. Fig. 16.21 shows a scale drawing of a lawn, to the scale of 1 cm per 10 m. A goat is tethered at point X, on a leash of length 15 m. Shade on the diagram the region which the goat can graze.

Fig. 16.21

3. The scale of Fig. 16.22 is 1 cm to 100 km. A, B and C are three radio transmitting stations. They each have a range of 300 km. Shade on the map the region within which all three stations can be received.

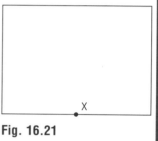

Fig. 16.22

4. Fig. 16.23 is a map of the entrance to a harbour, drawn to a scale of 1 cm : 2 km. To enter the harbour, a ship must steer so that it is an equal distance from points X and Y. Plot the course of the ship on the diagram.

Fig. 16.23

Revision checklist

This chapter has revised:

- 16.1 Measuring lengths and angles. Constructing triangles. ❑
- 16.2 Scales of maps. Distances and bearings of points. ❑
- 16.3 The locus of a point moving according to a given rule. ❑

CHAPTER 17 Similarity and Pythagoras

Chapter key points
- When using a scale diagram, be careful with scales in which the units are different, e.g. when the real object is measured in metres and the diagram in centimetres.
- The order of letters is important in congruence and similarity. If $\triangle ABC$ is congruent to $\triangle DEF$, then $\hat{A} = \hat{D}$ etc., $AB = DE$ etc. $\triangle ABC$ is not necessarily congruent to $\triangle EFD$.
- Pythagoras' theorem holds only for right-angled triangles.
- The hypotenuse is the longest side of a right-angled triangle. It is a good idea to label it before using the equation $c^2 = a^2 + b^2$.
- Do the algebra in the correct order. You need to use the expression:

$$\sqrt{a^2 + b^2} \quad \text{or} \quad \sqrt{c^2 - a^2}$$

so square first, then add or subtract, then take the square root. Don't over-simplify:

$$\sqrt{a^2 + b^2} \neq a + b$$

17.1 Scale diagrams

Similar figures have exactly the same shape. In particular they have the same angles. The figures may not be the same size.

A **scale diagram** of a real object is similar to it, in this mathematical sense. The scale is the ratio between lengths in the diagram and lengths in the real object.

■ Example 17.1
A building is 20 m high and 30 m wide (Fig 17.1). A scale diagram of the building is 40 cm high.

a) What is the scale of the diagram?
b) What is the width of the diagram? Give your answer in cm.

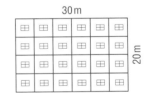

Fig. 17.1

Solution
a) The scale is 40 cm to 20 m. Convert the m to cm. The scale is 40 cm to 2000 cm. Divide both of these terms by 40.
   The scale is 1 : 50.
b) Divide the real width by the scale, obtaining 0.6 m. Convert to cm by multiplying by 100.
   The width of the diagram is 60 cm.

Key point
- Be careful with scales in which the units are different, e.g. when the real object is measured in metres and the diagram in centimetres.

SIMILARITY AND PYTHAGORAS

EXERCISE 17A

1. A model of an aircraft is in the scale 1 : 72. If the model is 25 cm long, what is the length of the real aircraft? Give your answer in m.
2. A model of a ship is 60 cm long. The real ship is 90 m long. What is the scale of the model? If the model is 20 cm wide, what is the width of the ship?
3. A diagram of a room is drawn to the scale 1 : 30.
    a) The width of the room is 3.6 m. What is its width on the diagram?
    b) The length of the room on the diagram is 15 cm. What is the length of the real room?
4. A rectangular field is 100 m by 150 m. A map of the field has the scale 1 : 500. Find the length of the field's sides on the map, and hence find the area of the field on the map.
5. A park is 2 km long. A map of the park is 80 cm long. What is the scale of the map? A lake in the park is a rectangle which is 500 m by 200 m. Find the area of the lake on the map.

17.2 Congruent and similar triangles

- **Congruent triangles** have exactly the same shape *and* size.
- **Similar triangles** have the same shape – they have exactly the same angles – but they may not be the same size.

Fig. 17.2 shows two similar triangles, ABC and DEF. The sides of △ABC are k times the sides of △DEF, where k is constant. Hence:

$$\frac{AB}{DE} = \frac{BC}{EF} = \frac{AC}{DF} = k$$

The ratio of sides within the triangles is the same for both triangles:

$$\frac{AB}{BC} = \frac{DE}{EF} \qquad \frac{AB}{AC} = \frac{DE}{DF} \qquad \frac{AC}{BC} = \frac{DF}{EF}$$

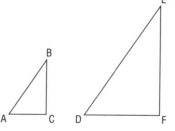

Fig. 17.2

Similar triangles have the same angles:

$$\hat{A} = \hat{D} \qquad \hat{B} = \hat{E} \qquad \hat{C} = \hat{F}$$

Note that the order of the letters is important: if △ABC is similar to △DEF, $\hat{A} = \hat{D}$ etc.

■ Example 17.2

Seen from the Earth, the Sun and the Moon appear to be the same size. The approximate distances from the Earth to the Sun and the Moon are 150 000 000 km and 400 000 km respectively. The diameter of the Moon is 3500 km. Find the diameter of the Sun.

Solution

Fig. 17.3 shows the situation at the time of an eclipse. The triangles EAB and ECD are similar. Using the property of similar triangles:

$$\frac{CD}{AB} = \frac{EC}{EA}, \quad \text{i.e.} \quad \frac{CD}{3500} = \frac{150\,000\,000}{400\,000}$$

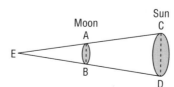

Fig. 17.3

CD = 3500 × 150 000 000 ÷ 400 000 = 1 300 000

The diameter of the Sun is approximately 1 300 000 km.

SIMILARITY AND PYTHAGORAS

■ Example 17.3

Fig. 17.4 shows a triangular prism ABCDEF. Find a pair of congruent triangles.

Solution

A triangular prism has constant cross-section. Hence the triangles at the end are congruent. Be sure to get the letters in the right order.
Triangles ABC and DEF are congruent.

Fig. 17.4

Key point

■ The order of letters is important in congruence and similarity. If △ABC is congruent to △DEF, then $\hat{A} = \hat{D}$ etc., AB = DE etc. △ABC is not necessarily congruent to △EFD.

EXERCISE 17B

1. In Fig. 17.5, X and Y are on the sides AB and AC of the triangle ABC, so that XY is parallel to BC. Write down a pair of similar triangles. AB = 15 m, AX = 5 m and XY = 4 m. Find BC.

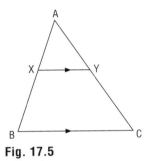

Fig. 17.5

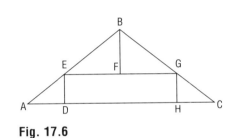

Fig. 17.6

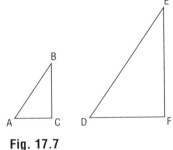

Fig. 17.7

2. Fig. 17.6 shows part of a roof truss. Write down as many similar triangles as you can.

3. In Fig. 17.7 the triangles ABC and DEF are similar. The diagram is not drawn to scale. The following distances are in cm.
   a) AB = 10, DE = 5 and AC = 7. Find DF.
   b) AB = 6, AC = 4 and DE = 18. Find DF.
   c) BC = 9, EF = 12 and DE = 8. Find AB.
   d) AB = $\frac{1}{2}$, DE = $\frac{3}{4}$ and BC = $\frac{1}{3}$. Find EF.

4. Fig. 17.8 is a simplified picture of a pair of shears. The blades are 30 cm long, and the handles are 25 cm long. If the points of the blades are 40 cm apart, how far apart are the ends of the handles?

5. A girder is laid on top of two supports, as shown in Fig. 17.9. The higher support is $2\frac{1}{2}$ m above the ground, and the lower support is 2 m above the ground. The distance from the foot of the girder to the foot of the lower support is 3 m. How far is the foot of the girder from the foot of the higher support?

6. Fig. 17.10 shows a roof truss. The width AB is 8 m, and the post CD is 1.6 m from A. If the height of the post is 0.8 m, find the height of the roof ridge.

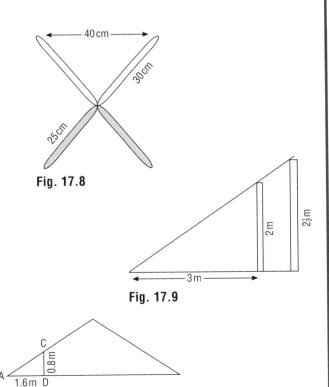

Fig. 17.8

Fig. 17.9

Fig. 17.10

© IT IS ILLEGAL TO PHOTOCOPY THIS PAGE

7 Fig. 17.11 shows a rectangle ABCD. It is cut along AC. Write down a pair of congruent triangles.

8 Fig. 17.12 shows an isosceles triangle. Show how to cut it into a pair of congruent triangles.

9 The dots of Fig. 17.13 are 1 cm apart.
   a) Find Z on the diagram, if triangles XYZ and ABC are congruent.
   b) Find N on the diagram, if triangles LMN and ABC are similar.

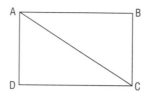
Fig. 17.11

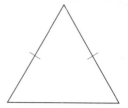
Fig. 17.12

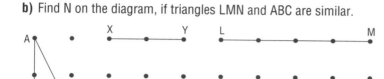
Fig. 17.13

17.3 Pythagoras' theorem

If you know two sides of a right-angled triangle then you can find the third side using **Pythagoras' theorem**.

Fig. 17.14 shows a right-angled triangle. The **hypotenuse** is the longest side, which is the side farthest from the right-angle. In the diagram c is the length of the hypotenuse. The theorem says:

$$c^2 = a^2 + b^2$$

In words:

The square of the hypotenuse is equal to the sum of the squares of the other two sides.

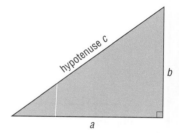
Fig. 17.14

■ Example 17.4

Fig. 17.15 shows a triangle which is right-angled at C, i.e. $\hat{C} = 90°$. CB = 7 cm and AC = 9 cm. Find BA.

Solution
Using the letters of Fig. 17.14, BA = c cm, AC = a cm, CB = b cm.
The theorem gives:
$c^2 = a^2 + b^2 = 9^2 + 7^2$
$c^2 = 81 + 49 = 130$, hence $c = \sqrt{130} = 11.4$
BA = 11.4 cm

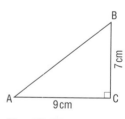
Fig. 17.15

■ Example 17.5

A ship sails for 100 km in a North-east direction. It ends up 70 km North of its starting point. How far East is it from the starting point?

Solution
Fig. 17.16 shows the journey. The hypotenuse is 100 km, and the vertical side is 70 km. Let the horizontal side be h km. From the theorem:
$100^2 = 70^2 + h^2$
$h = 100^2 - 70^2 = 10\,000 - 4900 = 5100$
Hence $h = \sqrt{5100} = 71.4$
It is 71.4 km East of the starting point.

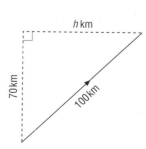
Fig. 17.16

SIMILARITY AND PYTHAGORAS

Key points
- Pythagoras' theorem holds only for right-angled triangles.
- The hypotenuse is the longest side of a right-angled triangle. It is a good idea to label it before using the equation $c^2 = a^2 + b^2$.
- Do the algebra in the correct order. You need to use the expression:

$$\sqrt{a^2 + b^2} \quad \text{or} \quad \sqrt{c^2 - a^2}$$

so square first, then add or subtract, then take the square root. Don't over-simplify:

$$\sqrt{a^2 + b^2} \neq a + b$$

EXERCISE 17C

1 For the triangles of Fig. 17.17, find the unknown sides. Lengths are in cm.

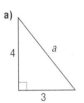

Fig. 17.17

2 A plane flies in a straight line, so that it is 250 miles North and 350 miles East of its starting point. How far has it flown?

3 Fig. 17.18 shows a kite being flown on level ground. The kite is 65 feet high, and the horizontal distance between the kite and the base of the string is 43 feet. How long is the string?

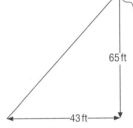

Fig. 17.18

4 The **gradient** of a road is the ratio of the vertical rise to the horizontal run. A steep mountain path has a gradient of 1 in 5. I walk up it so that I have travelled 30 m horizontally.

   a) How far have I risen?
   b) How far have I walked along the path?

5 A telegraph pole of length 5 m leans at an angle, so that its top is 4.2 m above the ground (Fig. 17.19). When the sun is vertically overhead, what is the length of the shadow of the pole?

6 A demolition ball is on the end of a 7 m chain. The ball is pulled aside 1.8 m to the left (Fig. 17.20).

   a) What is the vertical distance between the ball and the top of the chain?
   b) What height has the ball been raised through?

Fig. 17.19

Fig. 17.20

7 A ladder of length 3 m leans against a wall. The top of the ladder is 2.4 m above the ground.
How far is the foot of the ladder from the base of the wall?

8 Below are given the sides of two triangles. Is either triangle right-angled?

   a) 8 cm, 9 cm and 12 cm
   b) 15 m, 8 m and 17 m

9 Fig. 17.21 shows the side view of a roof. The sloping sections are 7 m long. The distance between the walls is 12 m. How high is the top of the roof above the walls?

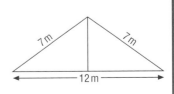

Fig. 17.21

10 Fig. 17.22 shows the cross-section of a log of wood. The log is a cylinder of radius 12 cm. The bottom 3 cm is to be cut away, as shown by the dotted line. What is the width of the cut?

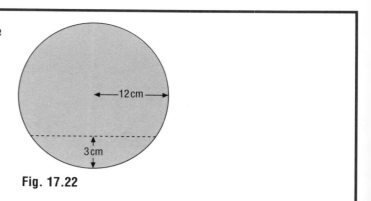

Fig. 17.22

Revision checklist

This chapter has revised:

17.1 Scale diagrams, in which there is a constant ratio between the diagram and the real object. ☐

17.2 Similar triangles and congruent triangles. ☐

17.3 Pythagoras' theorem, relating the sides of a right-angled triangle. ☐

CHAPTER 18 Trigonometry

Chapter key points
- The trig. ratios apply only to right-angled triangles. Don't use them for triangles without a right angle.
- Use the correct ratio. Until you are practised, label the sides of the triangle with HYP, OPP and ADJ.
- Make sure you know how your calculator works, i.e. whether you enter the angle first or the trig. function first.
- If you are asked to give your answer correct to 3 decimal places, work to 4 or more decimal places before giving the final answer.
- Angles can be measured in degrees, radians or grads. Put your calculator in degree mode. There should be a little 'D' or 'deg' on the display.
- Make sure you know when to use sin, cos or tan, and when to use $\sin^{-1}$, $\cos^{-1}$ or $\tan^{-1}$. Use the functions to find the sides, and the inverse functions to find the angles.

Suppose you have a right-angled triangle. If you know one side and one other angle, you can find the other sides. If you know two sides, you can find the other angles.

18.1 Finding the side

The triangle of Fig. 18.1 contains a right angle and an angle $P°$. Its sides are labelled as follows.

- Hypotenuse (HYP) – the longest side.
- Opposite (OPP) – the side opposite $P°$.
- Adjacent (ADJ) – the side next to $P°$.

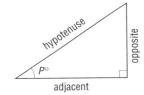

Fig. 18.1

The **trigonometric ratios** are ratios between the sides. They are functions of the angle $P°$.

$$\sin P° = \frac{\text{OPP}}{\text{HYP}} \quad \text{(the \textbf{sine} of } P°\text{)}$$

$$\cos P° = \frac{\text{ADJ}}{\text{HYP}} \quad \text{(the \textbf{cosine} of } P°\text{)}$$

$$\tan P° = \frac{\text{OPP}}{\text{ADJ}} \quad \text{(the \textbf{tangent} of } P°\text{)}$$

You can use the 'word' SOHCAHTOA to help you remember:

Sine is **O**pp over **H**yp, **C**os is **A**dj over **H**yp, **T**an is **O**pp over **A**dj

Use of a calculator

On some calculators the angle is entered first, on others the function is entered first. So the sequence for sin 30° might be:

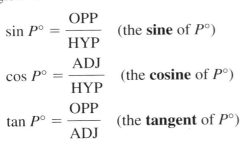

The answer 0.5 will appear.

TRIGONOMETRY

Angles of elevation and depression

Looking upwards, the **angle of elevation** is the angle between your line of vision and the horizontal.

Looking downwards, **the angle of depression** is the angle between your line of vision and the horizontal.

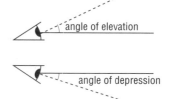

Fig. 18.2

■ Example 18.1

Fig. 18.3 shows a triangle ABC, for which $\hat{A} = 90°$, $\hat{B} = 48°$, and BC = 12 cm. Find AB, giving your answer correct to 3 significant figures.

Solution

First label the sides, by comparing with Fig. 18.1.
The HYP side is BC, the OPP side is AC, the ADJ side is AB.
We know the HYP and we want to know the ADJ. The function that links HYP and ADJ is cosine:

$$\cos 48° = \frac{\text{ADJ}}{\text{HYP}} = \frac{AB}{12}$$ Use a calculator to find that $\cos 48° = 0.6991$

This gives $0.6691 = \frac{AB}{12}$, hence $AB = 12 \times 0.6691 = 8.029$

AB = 8.03 cm to 3 significant figures

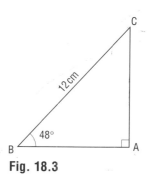

Fig. 18.3

■ Example 18.2

A boat is 1500 m out to sea. From the boat, the angle of elevation of the top of a cliff is 7°. How high is the cliff?

Solution

Let the height of the cliff be h m. In the triangle of Fig. 18.4, the sides are as follows.

height of cliff: OPP = h m
distance out to sea: ADJ = 1500 m
distance from cliff-top to ship: HYP

Fig. 18.4

We know the ADJ and we want to find the OPP. The tan function links OPP and ADJ:

$$\tan 7° = \frac{h}{1500} \quad h = 1500 \times \tan 7° = 184$$

The height of the cliff is 184 m.

Key points

- The trig. ratios apply only to right-angled triangles. Don't use them for triangles without a right angle.
- Use the correct ratio. Until you are practised, label the sides of the triangle with HYP, OPP and ADJ. You know one side and you want to know another side. Use the ratio that connects these two sides.
- Make sure you know how your calculator works, i.e. whether you enter the angle first or the trig. function first.
- If you are asked to give your answer correct to 3 decimal places, work to 4 or more decimal places before giving the final answer.
- Angles can be measured in degrees, radians or grads. Put your calculator in degree mode. There should be a little 'D' or 'deg' on the display.

EXERCISE 18A

1 In the triangles of Fig. 18.5, find the unknown sides *a*, *b* and *c*. Lengths are in cm.

a) 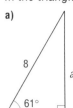   b) 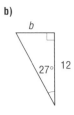   c)

Fig. 18.5

2 In triangle ABC, $\hat{A} = 90°$, $\hat{B} = 58°$, and BC = 9 cm. Find AB and CA.

3 Tom wants to find the height of a tree. He walks back 100 m from its base. From his feet, the angle of elevation of the top of the tree is 19° (Fig. 18.6). What is the height of the tree?

4 A straight beam of wood, of length $2\frac{1}{2}$ m, leans against a wall at an angle of 25° to the horizontal (Fig. 18.7). How high up the wall does the beam reach?

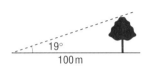
Fig. 18.6

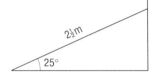
Fig. 18.7

5 Donna stands 40 m from the base of a tower (Fig. 18.8). The angle of elevation of the top of the tower is 42°. Given that her eyes are 1.4 m above the ground, how high is the tower?

6 A rocket is fired, so that it travels in a straight line making 65° with the horizontal. Its speed is 60 m/sec.

a) How far has it travelled after 10 seconds?
b) How high is it after 10 seconds?

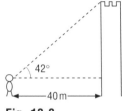
Fig. 18.8

7 Ivan wants to find the width of a river. He stands on one bank, and sees a tree directly opposite. He now walks 50 m along the bank. The line from him to the tree now makes 53° with the bank (Fig. 18.9). How wide is the river?

8 A ship sails from A to B, a distance of 650 km on a bearing of 079°. (Recall that bearings are measured clockwise from North.)

a) How far North is B from A?
b) How far East is B from A?

9 A railway track slopes at 2° to the horizontal. If a train travels 300 m up the track, by how much has it risen?

10 A flagpole stands vertically. When the sun is at an angle of elevation of 27°, the length of the shadow is 28 ft (Fig. 18.10). How high is the flagpole?

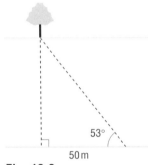
Fig. 18.9

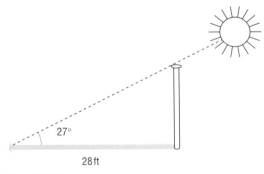
Fig. 18.10

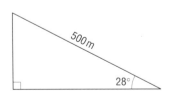
Fig. 18.11

11 Fig. 18.11 shows a triangular field in which the longest side is 500 m. Find the other sides and hence find the area of the field.

18.2 Finding the angle

In Exercise 18A you used sin, cos or tan to find a side of a right-angled triangle, given an angle and another side. You need to use the **inverse functions** $\sin^{-1}$, $\cos^{-1}$ or $\tan^{-1}$ to find an *angle* of a right-angled triangle, given two of the sides:

$$P° = \sin^{-1}\frac{\text{OPP}}{\text{HYP}} = \cos^{-1}\frac{\text{ADJ}}{\text{HYP}} = \tan^{-1}\frac{\text{OPP}}{\text{ADJ}}$$

Use of a calculator

On some calculators the ratio is pressed first, on others the inverse function button is pressed first. The sequence for $\cos^{-1} 0.5$ might be:

| . | 5 | inv | cos | or | inv | cos | . | 5 | = |

The answer 60 (deg) will appear.

■ Example 18.3

Fig. 18.12 shows $\triangle ABC$, for which $\hat{B} = 90°$, $AC = 8$ m and $AB = 3$ m. Find $\hat{C}$, giving your answer correct to the nearest degree.

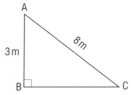

Fig. 18.12

Solution
We want to find $\hat{C}$. Label the sides of the triangle:
HYP = AC = 8 m, OPP = AB = 3 m, ADJ = BC
We know HYP and OPP. The function that links HYP and OPP is sine:

$$\sin \hat{C} = \frac{\text{OPP}}{\text{HYP}} = \frac{3}{8}$$

Hence $\hat{C} = \sin^{-1}\frac{3}{8} = 22°$

■ Example 18.4

A man stands on top of a 80 m high cliff, looking at a boat which is 350 m out to sea. What is the angle of depression of the boat from the top of the cliff?

Solution
Fig. 18.13 shows the situation. The angle of depression is $D°$. By alternate angles, the angle of elevation of the cliff from the boat is also $D°$. Label the triangle:

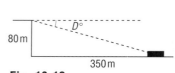

Fig. 18.13

HYP = distance to ship
OPP = height of cliff = 80 m
ADJ = distance out to sea = 350 m

We know OPP and ADJ. The function linking OPP and ADJ is tangent:

$$\tan D° = \frac{\text{OPP}}{\text{ADJ}} = \frac{80}{350}$$

Hence $D° = \tan^{-1}\frac{80}{350} = 12.9°$

The angle of depression is 12.9°.

Key point

- Make sure you know when to use sin, cos or tan, and when to use $\sin^{-1}$, $\cos^{-1}$ or $\tan^{-1}$. Use the functions to find the sides, and the inverse functions to find the angles.

EXERCISE 18B

1. In the triangles shown in Fig. 18.14, find the angles $a°$, $b°$ and $c°$. Lengths are in m. Give your answers correct to the nearest degree.

   a) 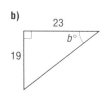 b)   c)

   Fig. 18.14

2. In the triangle ABC, $\hat{A} = 90°$, BC = 29 cm and AB = 19 cm. Find the angles $\hat{B}$ and $\hat{C}$.

3. The string of a kite is 18.6 m long. It is flown at a height of 12.7 m (Fig. 18.15). What angle does the string make with the horizontal?

   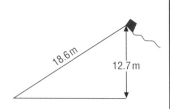
   Fig. 18.15

4. A flagpole of length 14 feet leans at an angle (Fig 18.16). When the sun is vertically overhead, the length of the shadow of the flagpole is 5 feet. What is the angle between the flagpole and the vertical?

5. A ladder of length $3\tfrac{1}{2}$ m leans against a wall (Fig 18.17). The ladder reaches 3 m up the wall. What is the angle between the ladder and the wall?

6. The road sign of Fig. 18.18 shows the gradient of a road. What is the angle of slope of the road?

7. A plane flies so that it is 400 miles North and 500 miles East of its starting point. What is the bearing of the plane from its starting point?

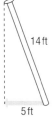

Fig. 18.16

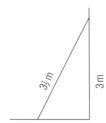

Fig. 18.17

Fig. 18.18

Revision checklist

This chapter has revised:

18.1 The three functions sin, cos and tan, as the ratios of the sides of a right-angled triangle. Using the ratios to find the sides of a right-angled triangle, given an angle and one side. ☐

18.2 The inverse functions $\sin^{-1}$, $\cos^{-1}$ and $\tan^{-1}$. Using the inverse functions to find the angles of a right-angled triangle, given two of the sides. ☐

CHAPTER 19 Transformations

> **Chapter key points**
> - Make sure you know the different types of transformation:
>   - a translation changes only the position of a shape
>   - an enlargement changes the size of a shape.
> - To distinguish between rotations and reflections, use the following. Suppose a shape is lettered clockwise. After rotation it is still lettered clockwise. After a reflection it will be lettered anti-clockwise.
> - In an enlargement, distances are measured from the centre of enlargement. If the scale factor is 2, distances from the centre X are doubled, i.e. XA' = 2 × XA. Don't measure the distance from the original position, i.e. don't make AA' = 2 × XA. This would give a scale factor of 3.
> - Any shape, however irregular, is the same after a rotation of 360°. This does not mean it has rotational symmetry of order 1. It has no rotational symmetry.

A **transformation** changes the position of points and shapes in various ways.

19.1 Translations and reflections

A **translation** shifts all points and shapes in a fixed direction. If a translation shifts x to the right and y up, it is represented by the **vector**:

$$\begin{pmatrix} x \\ y \end{pmatrix}$$

A **reflection** moves each point to a point the same distance on the other side of a fixed line.

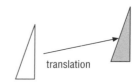

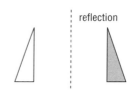

Fig. 19.1

■ Example 19.1

A triangle is shown on the grid of Fig. 19.2. Show its position after it has been:

a) reflected in the dotted line
b) translated 2 to the right and 1 up.

Solution
The effect of these operations is shown in Fig. 19.3.

Fig. 19.2

■ Example 19.2

In Fig. 19.4 the triangle A is translated to the triangle B. Write down the vector of this translation.

Solution
The translation shifts 2 to the right and 1 down.

The translation is $\begin{pmatrix} 2 \\ -1 \end{pmatrix}$.

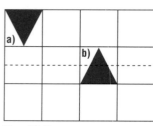

Fig. 19.3

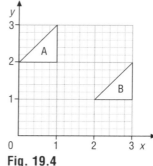

Fig. 19.4

TRANSFORMATIONS

EXERCISE 19A

1. Reflect the shape in Fig. 19.5 in the mirror line AB.
2. In Fig. 19.6, translate the triangle 2 to the right and 3 up.

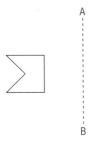

Fig. 19.5

Fig. 19.6

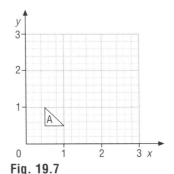

Fig. 19.7

3. In Fig. 19.7, apply the vector translation $\begin{pmatrix} 1 \\ 2 \end{pmatrix}$ to the triangle A.

   Find the coordinates of the vertices of the new triangle.

4. A translation takes the point (1, 3) to (2, −1). Write down the vector which represents this translation. Where is (4, 2) taken by this translation?

5. Complete the shape in Fig. 19.8 by reflection in the dotted line.

6. The triangle A in Fig. 19.9 is reflected in the line $x = 2$, to give triangle B.
   Triangle B is then reflected in the line $x = 4$, to give triangle C.
   Draw B and C on the diagram. A translation will take A directly to C. Describe the translation.

7. The triangle in Fig. 19.10 is reflected in the line $y = x$ (shown broken in the diagram). Draw the reflected triangle, and write down the coordinates of its vertices.

Fig. 19.8

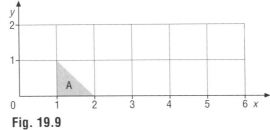

Fig. 19.9

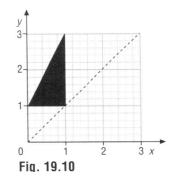

Fig. 19.10

19.2 Rotations and enlargements

A **rotation** turns shapes through a fixed angle about a point. An **enlargement** increases the size of shapes by a constant number, called the **scale factor**.

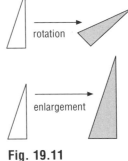

Fig. 19.11

■ Example 19.3

On the grid of Fig. 19.12, rotate the triangle ABC clockwise through 90°.

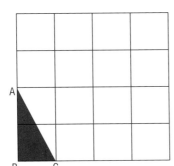

Fig. 19.12

Solution
The vertical side AB becomes horizontal, and the horizontal line BC becomes vertical. The result is shown in Fig. 19.13.

■ Example 19.4
Enlarge the triangle ABC of Fig. 19.12 by a scale factor of 2.

Solution
AB has length 2 units: enlarge to 4 units. BC has length 1 unit: enlarge to 2 units. The result is shown in Fig. 19.14.

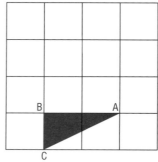

Fig. 19.13

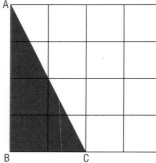

Fig. 19.14

Key point
■ Make sure you know the different types of transformation:
- a translation changes only the position of a shape
- an enlargement changes the size of a shape

■ To distinguish between rotations and reflections, use the following. Suppose a shape is lettered clockwise. After rotation it is still lettered clockwise. After a reflection it will be lettered anti-clockwise.

EXERCISE 19B

1. Rotate the triangle ABC in Fig. 19.15 clockwise through 180°. Sketch the result.
2. Rotate the triangle ABC in Fig. 19.16 clockwise through 90°. Sketch the result.
3. In Fig. 19.17, the triangle T has been transformed to T'. Through what angle has it been rotated?
4. Enlarge the triangle in Fig. 19.18 by a scale factor of 3.
5. In Fig. 19.19, the triangle S has been transformed to S'. What is the scale factor of the enlargement?

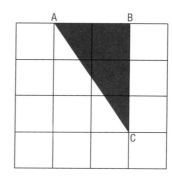

Fig. 19.15

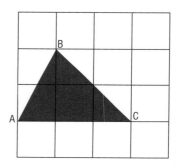

Fig. 19.16

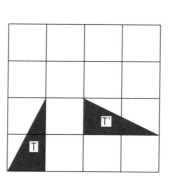

Fig. 19.17

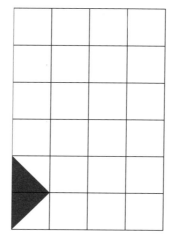

Fig. 19.18

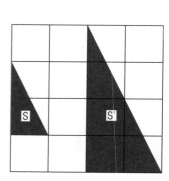

Fig. 19.19

TRANSFORMATIONS

19.3 Centres of rotation and enlargement

The **centre of rotation** is the point that remains fixed under a rotation. In Fig. 19.20, X is the centre of rotation. Notice that XA has been rotated 90° clockwise to XA'.

The **centre of enlargement** is the point that remains fixed under an enlargement. Distances from the centre of enlargement are multiplied by the scale factor λ. The scale factor λ can be less than 1. In this case the shapes become smaller.

In Fig. 19.21, X is the centre of enlargement. The scale factor is 2. Notice that

$$XA' = 2 \times XA$$

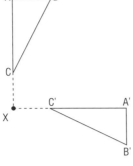

Fig. 19.20

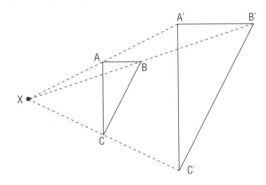

Fig. 19.21

To find the centre of enlargement, join up the corresponding points on the original and the enlarged shape, and extend. Note that in Fig. 19.21 AA', BB' and CC' all meet at X.

■ Example 19.5

The triangle ABC of Fig. 19.22 is rotated through 180° about X. Draw the new triangle.

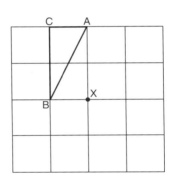

Fig. 19.22

Solution
Point A is 2 units above X. After rotation, it is 2 units below X.
Point B is 1 unit to the left of X. After rotation, it is 1 unit to the right of X.
Point C is 1 unit to the left of A, hence after rotation it is 1 unit to the right of the new position of A.
The result is shown in Fig. 19.23.

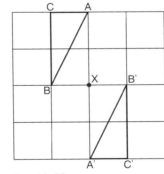

Fig. 19.23

■ Example 19.6

The triangle ABC of Fig. 19.24 is enlarged by a scale factor of 2, with centre of enlargement at X. Draw the new triangle.

Solution
Point A is 2 units to the right of X.
After enlargement, it is 4 units to the right of X.
Point C is 1 unit to the right of X.
After enlargement, it is 2 units to the right of X.
Point B is 1 unit to the right and 2 units down from X. After enlargement, it is 2 units to the right and 4 units down from X.
The enlarged triangle is shown in Fig. 19.25.

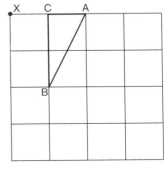

Fig. 19.24

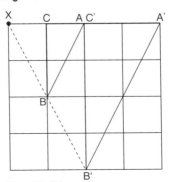

Fig. 19.25

© IT IS ILLEGAL TO PHOTOCOPY THIS PAGE

Example 19.7

In Fig. 19.26, the black triangle has been enlarged to the grey triangle. Find the scale factor and the centre of enlargement.

Solution

The vertical side of the black triangle is 1 unit, and the vertical side of the white triangle is 3 units.
The scale factor is 3.
Join up corresponding points, as shown in Fig. 19.27.
The dotted lines meet at (0, 3).
The centre of enlargement is (0, 3).

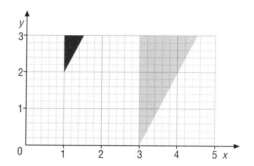

Fig. 19.26   Fig. 19.27

Key point

- In an enlargement, distances are measured from the centre of enlargement. If the scale factor is 2, distances from the centre X are doubled, i.e. XA' = 2 × XA. Don't measure the distance from the original positions, i.e. don't make AA' = 2 × XA. This would give a scale factor of 3.

EXERCISE 19C

1. On the grid of Fig. 19.28, rotate the triangle by 180° about X.

2. Rotate the triangle in Fig. 19.29 anti-clockwise through 90° about the origin O. Write down the coordinates of the new vertices.

3. Enlarge the triangle in Fig. 19.30 by a factor of 2, with centre of enlargement (0, 2). Write down the coordinates of the new vertices.

4. Enlarge the triangle in Fig. 19.31 by a factor of $\frac{2}{3}$, with centre of enlargement (3, 0). Write down the coordinates of the new vertices.

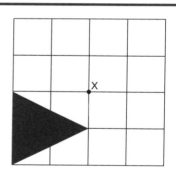

Fig. 19.28

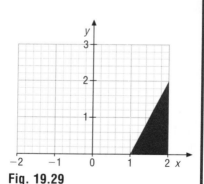

Fig. 19.29

5. In Fig. 19.32 the white triangle has been enlarged to the coloured triangle. Find the scale factor and the centre of enlargement.

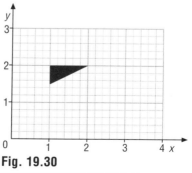

Fig. 19.30

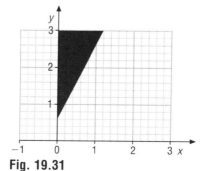

Fig. 19.31

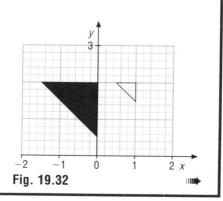

Fig. 19.32

6   Fig. 19.33 shows the view from above of part of a room. The rectangle represents a wardrobe, which is to be moved from position A to position B. Describe rotations which will achieve this.

7   The triangle in Fig. 19.34 is reflected in the *y*-axis. Draw the reflected triangle. The result is then reflected in the line *y* = *x* (shown as a broken line). Draw this triangle. The two reflections are equivalent to a rotation. Find the angle and centre of the rotation.

Fig. 19.33

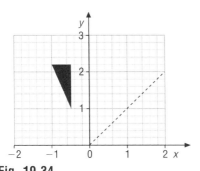

Fig. 19.34

19.4 Symmetry

Some important terms connected with symmetry are defined below (Fig. 19.35).

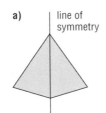

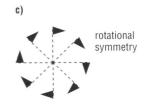

Fig. 19.35

Line symmetry   A flat shape is unchanged after reflection in a line of symmetry.
Plane of symmetry   A solid object is unchanged after reflection in a plane of symmetry.
Rotational symmetry   If a shape is unchanged after a rotation (of less than 360°) about a point, it has rotational symmetry.
Order of rotational symmetry This is the number of times it fits its own outline in a complete turn.

■ Example 19.8

Describe the line and rotational symmetries of the symbol in Fig. 19.36.

Fig. 19.36

Solution

The symbol will remain unchanged if it is reflected about any of the three spokes. See Fig. 19.37.
There are three lines of symmetry.
It will also remain unchanged if it is rotated through a third of a circle about the centre.
The symbol has rotational symmetry of order 3.

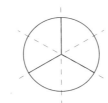

Fig. 19.37

■ Example 19.9

Find a plane of symmetry of the cube in Fig. 19.38

Solution

The cube is identical on either side of the vertical plane through AC.
ACGE is a plane of symmetry.
Note. There are many other planes of symmetry.

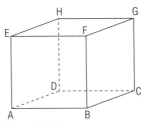

Fig. 19.38

Key point

■ Any shape, however irregular, is the same after a rotation of 360°. This does not mean it has rotational symmetry of order 1. It has no rotational symmetry.

EXERCISE 19D

1 How many lines of symmetry has:

   a) a square   **b)** a rectangle   **c)** a kite   **d)** a rhombus?

2 Do any of the shapes of Fig. 19.39 have symmetry? If they do, state the number of lines of symmetry and the order of rotational symmetry.

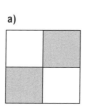
a)
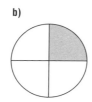
b)

c)

Fig. 19.39

Fig. 19.40

3 Fig. 19.40 shows a regular pentagon. How many lines of symmetry does it have? What is the order of its rotational symmetry?

4 Describe the symmetry of the following letters.

   H   N   I   X   T   Z

5 Complete the shapes of Fig. 19.41 so that they have symmetry about the dotted line.

Fig. 19.41

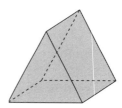

Fig. 19.42

6 The prism of Fig. 19.42 has a cross-section which is an equilateral triangle. How many planes of symmetry does the prism have?

7 How many planes of symmetry has a cube?

Revision checklist

This chapter has revised:

19.1 Translations, which shift points in a fixed direction. Reflections, which move points to the other side of a mirror line. ❏

19.2 Rotations, which turn points through a fixed angle. Enlargements, which increase lengths by a fixed factor. ❏

19.3 The centre of rotation, which remains stationary under a rotation. The centre of enlargement, which remains stationary under an enlargement. Finding the centre of enlargement. ❏

19.4 Lines of symmetry of a flat shape, and planes of symmetry of solid shapes. Rotational symmetry and order of rotational symmetry. ❏

Mixed exercise 3

1. Find the unknown angle $x°$ in Fig. M3.1.

2. Every morning Janet goes to swim in a lake. The length of the lake is 90 m. She swims to the end and back twice.

    a) How far does she swim?
    b) She takes 15 minutes for her swim. What is her average speed, in metres per second?
    c) 1 metre is approximately equal to 1.083 yards. How many yards does she swim every day?

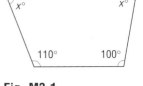

Fig. M3.1

3. Fig. M3.2 shows a cuboid, which is 4 cm by 3 cm by $4\frac{1}{2}$ cm.

    a) How many vertices does the cuboid have?
    b) What is the volume of the cuboid?
    c) Draw an accurate net of the cuboid. One face has already been drawn for you in Fig. M3.3.

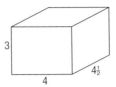

Fig. M3.2

4. Fig. M3.4, which is not to scale, shows a triangle ABC for which AB = 4 cm, BC = 5 cm and CA = $3\frac{1}{2}$ cm. Make an accurate construction of △ABC. Measure the angle $\hat{B}$.

5. In the following sentences, insert a word from this list:

    kite   trapezium   rectangle   rhombus

    a) A ... has opposite sides equal, and equal diagonals.
    b) A ... has two pairs of equal sides, and one line of symmetry.
    c) A ... has two lines of symmetry, and its diagonals meet at 90°.
    d) A quadrilateral with only two parallel sides is a ....

Fig. M3.3

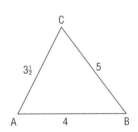

Fig. M3.4

6. On the isometric paper of Fig. M3.5 draw a cuboid which is 1 unit by 1 unit by 3 units.

7. Fig. M3.6 shows a triangle S on a grid. The lines of the grid are 1 unit apart.

    a) S is translated 3 units to the right and 3 units up. Draw the new triangle.
    b) S is reflected in the dotted line. Draw the new triangle.

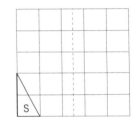

Fig. M3.5   **Fig. M3.6**

8. A pond is a circle of radius 70 m.

    a) What is the circumference of the pond?
    b) What is the area of the pond?

9. Fig. M3.7 shows a regular polygon ABCDEF. The diagonals AD, BE and CF meet at X.

    a) What is the name of this polygon?
    b) What is the order of rotational symmetry of the polygon?
    c) The triangle ABX can be rotated to the triangle CDX.
        i) What is the angle of rotation?
        ii) What is the centre of rotation?

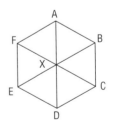

Fig. M3.7

10. Fig. M3.8 is a scale diagram of a rectangular room. It is 4 cm by 5 cm.
    The shorter side of the real room is 8 m long.

    a) What is the scale of the diagram?
    b) What is the length of the longer side of the real room?

11. Fig. M3.9 shows several shapes. Find two pairs of shapes that are congruent.

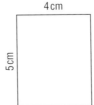

Fig. M3.8

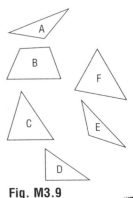

Fig. M3.9

12 Find the area and perimeter of the shape of Fig. M3.10. Lengths are in cm.

13 Fig. M3.11 shows a right-angled triangle. By drawing at least five more triangles, show how this triangle can be used to tessellate a surface.

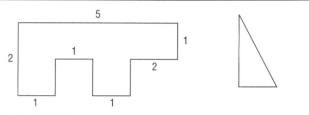

Fig. M3.10   Fig. M3.11

The following questions are suitable for Intermediate Level.

14 Matthew walks up a straight slope. He walks 80 m, rising through a vertical distance of 25 m.

   a) How far has he gone horizontally?
   b) What is the angle of the slope?

15 In Fig. M3.12, the triangle S has been enlarged to S'.

   a) What is the scale factor of the enlargement?
   b) What is the centre of enlargement?

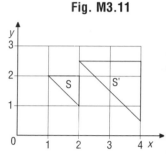

Fig. M3.12

16 A can of paint is a cylinder with base radius 8 cm.

   a) If the height is 17 cm, what is the volume?
   b) If the volume is 3000 cm³, what is the height?

17 a) In Fig. M3.13, AB is a diameter of the circle, and C is a point on its circumference.
   The angle CÂB is 39°. Find AĈB and CB̂A.
   b) In Fig. M3.14, AB is a diameter of the circle. The chord CD crosses AB at right angles, at X.

   i) What can you say about CX and DX?
   ii) What can you say about △ACX and △ADX?

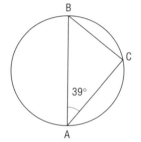 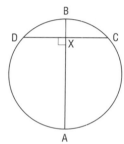

Fig. M3.13   Fig. M3.14

18 In Fig. M3.15, AB is parallel to CD. Write down a pair of similar triangles. If AB = 4 cm, AX = 6 cm and CD = 6 cm, find XD.

19 A formula for the volume of an object is $V = \pi r(rh + r)$, where r and h are lengths.

   a) Explain why the formula must be incorrect.
   b) Suggest a possible correct formula for the volume.

20 Fig. M3.16 is a square-based pyramid, in which the vertex V is immediately above the base ABCD.

   a) How many planes of symmetry does the solid have?
   b) How many axes of symmetry does the solid have?

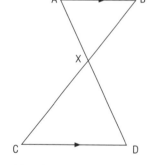 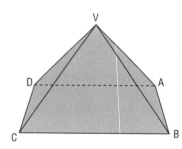

Fig. M3.15   Fig. M3.16

21 Find the unknown angle $x°$ in Fig. M3.17.

22 Fig. M3.18 shows a map of a region of sea. The scale is 1 cm for 20 km. A ship leaves port at P and sails on a bearing of 048°.

   a) Plot the ship's course on the map.

   There is a lighthouse at L. The signal from the lighthouse can be seen for 30 km.

   b) On the map, indicate the region within which the signal can be seen.
   c) Will the ship be able to see the signal? If so, find the distance from P when it first does so.

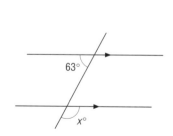 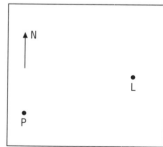

Fig. M3.17   Fig. M3.18

Section 4
HANDLING DATA

CHAPTER 20 Collecting data

Chapter key points
- A questionnaire must obtain the information it is intended for. It must be unbiased, precise, simple to answer and to complete.
- On a tally sheet each |||| symbol represents five items, not four.

Data are items of information. They may be numbers, or names, or colours. You might collect data for investigations.

20.1 Reading tables

Example 20.1

The table below shows the pets in four households.
a) Which households have a cat?
b) What pets are there in the Green household?

	Smith	Jones	Green	Brown
Cat		✓		✓
Dog	✓		✓	✓
Hamster	✓			✓
Gerbil			✓	

Solution
a) Look along the 'cat' row.
   The Jones and Brown households have a cat.
b) Look down the 'Green' column.
   The Green household has a dog and a gerbil.

EXERCISE 20A

1 The features of four types of motorcycle are shown in the table below.

	Disc brakes	Electric starter	Alarm	Rev. counter
Harakiri		✓		✓
Binbo				
Kanemochi	✓	✓	✓	✓
Anzen	✓		✓	

a) Which motorcycle has all the features?
b) Which motorcycle has none of the features?
c) Which types have an alarm?

2   The table below shows the foreign languages spoken by five people.

	Melanie	Jonathon	Jordan	Georgie	Alex
French	✓	✓	✓		✓
German		✓			✓
Spanish	✓	✓		✓	

a) Who speaks most foreign languages?
b) Which language is spoken by most people?

3   Four friends are taking A-levels. Alan is taking Maths and Physics, Belinda is taking Maths and Economics, Christine is taking Biology, Physics and Chemistry, and David is taking Maths, Economics and Chemistry. Show this information in a table.

4   Four people are talking about which items of electrical equipment they have. Eva and Francis have a computer, Eva, George and Harriet have a CD player, Francis and Harriet have a television, and all four have a radio. Show this information in a table.

5   The table below shows how a group of schoolchildren came to school.

	Car	Bus	Train	Cycle	Walking
Number	10	27	13	9	43

a) How many came by train?
b) How many children were there in total?

6   The results of an exam are given in the table below.

Grade	A	B	C	D	E	F	U
Number	29	59	217	84	88	33	19

a) How many candidates got below grade D?
b) How many took the exam in total?
c) Grades A, B and C are counted as passes. How many passed the exam?

7   The table below gives the numbers of second-hand cars of different ages and different price ranges.

Price (£)						
8000–9999	12	8	1	0	0	0
6000–7999	7	15	8	1	0	0
4000–5999	3	17	21	6	4	1
2000–3999	0	1	5	8	6	3
	1	2	3	4	5	6
	Age (years)					

a) What is the newest car I can buy with less than £4000?
b) What proportion of the cars cost at least £6000?

20.2 Questionnaires

A **questionnaire** finds out facts about people, or it finds their opinion on a topic. When designing a questionnaire, make sure that it obtains the correct information. It should be:

Unbiased   It should not influence the replier.
Precise   The replier should know exactly what the question means.
Easy   The replier should find the question easy to answer.
Complete   Every possible answer should be allowed for.

Finding opinions

The question might ask for an opinion on a topic. It makes it easier to reply if the various choices are presented for ticking, for example:

strongly approve   ☐
approve   ☐
neutral   ☐
disapprove   ✓
strongly disapprove   ☐

Finding facts

The question might ask for a factual answer, for example a number. Again, it makes it easy to reply if the replier just has to circle one of various choices, for example:

    0   1   ②   3   4   5+

■ Example 20.2

Rosa wants to find out to what extent people play the National Lottery. Her questionnaire includes the following question. Suggest three improvements, and write out a revised question.

How many Lottery entries do you make each Saturday? Circle one answer.

    1   2   3   4

Solution

There is no box for 0 entries. There is no box for more than 4 entries. People vary their number of entries from week to week, so it is better to be precise about which day is meant.
An improved version could be:
How many Lottery entries did you make last Saturday? Circle one answer.

    0   1   2   3   4+

Key point

■ A questionnaire must obtain the information it is intended for. It must be unbiased, precise, simple to answer and to complete.

COLLECTING DATA

EXERCISE 20B

1. Barbara wants to find out the popularity of video films. Her questionnaire includes the following question. Suggest improvements, and rewrite the question.

   How many videos did you hire over the past year?

      1   2   3   4   5+

2. Damian wants to find out how long people spend on their Maths homework. He asks people to answer the following question. Write out an improved version.

   How long do you spend on your Maths homework?

      0–2 hours   2–4 hours   4–6 hours   6–8 hours

3. An animal welfare group is seeking a ban on fishing for sport. Its questionnaire includes the following. Write out an improved question.

   Do you think the suffering caused by angling is justified?

4. A 'Eurosceptic' organisation submits a questionnaire with the following wording. Rewrite it.

   The European Courts are meddling with our internal justice system. Do you:

   strongly approve ☐
   approve ☐
   don't care ☐
   disapprove ☐
   strongly disapprove? ☐

5. Your Maths teacher is seeking your opinions on her teaching. She gives you a questionnaire which includes the following. Suggest how it could be improved.

   How do you rate my teaching? Circle one answer.

      1   2   3   4   5

6. A restaurateur wants to open a new Chinese restaurant in the High Street of a town. He needs to know how popular it will be. Design a suitable questionnaire to find out how often it would be used.

7. Design a questionnaire to find out how long people spend watching television.

8. Design a questionnaire to find out people's opinions on capital punishment.

20.3 Tallies and frequencies

When you are counting data, it often helps to make a **tally sheet**. The items are recorded by strokes, in groups of up to five:

   | = 1 || = 2 ||| = 3 |||| = 4 ||||̶ = 5

The numbers of data that make up the groups are the **frequencies**.
The frequencies can be listed in a table.

COLLECTING DATA

■ Example 20.3

The tally sheet below was used to record the traffic passing a house. Construct a frequency table to show the information.

Cars	Vans	Lorries	Buses	Motorbikes
ⅢⅠ	ⅢⅠ	ⅢⅠ	ⅠⅠ	ⅢⅠ
ⅢⅠ	Ⅰ	ⅢⅠ		ⅠⅠⅠ
ⅢⅠ		ⅠⅠⅠ		
ⅠⅠ				

Solution

Each bundle ⅢⅠ corresponds to five items. Hence there were 17 cars, 6 vans and so on. The frequency table is below.

Cars	Vans	Lorries	Buses	Motorbikes
17	6	13	2	8

Key point

■ Each ⅢⅠ symbol represents five items, not four.

EXERCISE 20C

1 A tally sheet was used to find the number of pets in households. Record the data in a frequency table.

Number of pets:	0	1	2	3	4+
	ⅢⅠ	ⅢⅠ	ⅠⅠⅠ	ⅠⅠⅠ	ⅠⅠⅠⅠ
	ⅠⅠⅠ	ⅢⅠ			

2 The tally chart below records where people went for their most recent foreign holiday. Convert it into a frequency table.

France	Italy	Spain	Greece	USA	Other
ⅢⅠ	ⅢⅠ	ⅢⅠ	ⅠⅠⅠ	ⅠⅠ	ⅢⅠ
ⅠⅠⅠ	Ⅰ	ⅢⅠ			ⅢⅠ
		ⅠⅠⅠ			ⅠⅠⅠⅠ

3 A die was rolled 40 times and the score written down. The results are below. Complete a tally sheet to record the data, and make a frequency table.

6 2 3 3 1 5 1 4 2 6 3 5 3 3 4 2 1 3 4 6
3 4 2 2 4 1 5 3 5 2 5 2 6 2 4 5 2 1 3 2

1

2

3

4

5

6

4 The ages of 30 people at a party were as follows. Fill in the ages on a tally sheet, then construct a frequency table.

16 25 31 16 18 20 38 35 32 23 25 28 19 20 21
22 18 27 17 16 34 22 29 17 26 21 17 33 20 32

Age	15–19	20–24	25–29	30–34	35–39

5 The grades in Maths GCSE obtained by a class of 25 pupils are given below. Construct a tally sheet and a frequency table.

B C C C B C D B C C C D X B C X B C C C B
D D C C

6 The colours of cars in a car park were as below. Construct a tally sheet and a frequency table.

red   blue   black   green   blue   blue   white   black   white   red   yellow
white   black   white   red   blue   black   white   white   blue   black

Revision checklist

This chapter has revised:

20.1 Obtaining information from tables. ❏
20.2 Preparing questionnaires to find information from people. ❏
20.3 Preparing tally sheets to record frequencies, and frequency tables. ❏

CHAPTER 21 Pictures of data

> **Chapter key points**
> - When constructing a pie chart, add up the angles to ensure that their sum is 360°.
> - If a bar chart displays continuous numerical information, the bars should touch each other. If the information is not continuous, the bars should not be touching.
> - A line of best fit through a scatter diagram goes through the 'middle' of the points. There should be about as many points above the line as below it. The line doesn't necessarily go through the first and last points, nor through the origin.

Many people understand information more easily if it is presented in a diagram. Here we discuss several sorts of statistical diagram.

21.1 Pictograms, bar charts, pie charts

A **pictogram** shows the number of items by symbols. Each symbol represents a fixed number of items.

If the items are divided into groups, a **bar chart** may be used. This consists of bars, whose lengths are proportional to the sizes of the groups. Alternatively, a **pie chart** may be used. This consists of sectors, whose angles are proportional to the sizes of the groups.

■ Example 21.1
The pictogram of Fig. 21.1 shows the population of Aceville. Each symbol represents 1 000 000 people. What is the population of Aceville? Bigtown has a population of 3 500 000. Draw a pictogram for Bigtown.

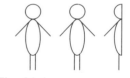

Fig. 21.1

Solution
There are $2\frac{1}{2}$ symbols. Multiply by 1 000 000.
The population of Aceville is 2 500 000.
For Bigtown, draw $3\frac{1}{2}$ symbols, as in Fig. 21.2.

Fig. 21.2

■ Example 21.2
An ice-cream van sells ice-cream in vanilla, chocolate, strawberry and peppermint flavours. The table below gives the numbers sold over an afternoon. Construct a bar chart on Fig. 21.3 to show the information.

Vanilla	Chocolate	Strawberry	Peppermint
12	10	8	6

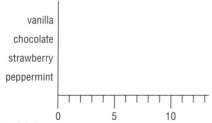

Fig. 21.3

Solution
There are 12 items in the 'Vanilla' bar. Draw a bar which is 12 units long. Do the same for the other flavours. The result is Fig. 21.4.

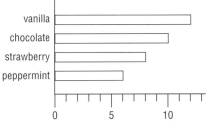

Fig. 21.4

■ Example 21.3
A group of 30 people were asked to state their favourite holiday activity. The pie chart of Fig. 21.5 shows the results.

a) 10 people like sight-seeing best. What is the angle in that sector of the pie?
b) The sector in the sunbathing sector has angle 96°. How many people like this best?

Fig. 21.5

Solution
a) 10 is a third of 30. Take a third of 360°.
   The angle in the sight-seeing sector is 120°.
b) Divide 96 by 360. Then multiply by the total number of people, i.e. by 30.

$$\frac{96}{360} \times 30 = 8$$

8 people like sunbathing best.

EXERCISE 21A

1. The daily production of a bakery is shown in the pictogram of Fig. 21.6, in which each picture represents 1000 loaves. How many loaves are made each day? Draw a pictogram for an increased daily production of 4500 loaves.

2. The pictogram of Fig. 21.7 shows the sales of pizza over six days. Each symbol represents 100 pizzas. Complete the table below to show the same information.

Day of week	Mon	Tue	Wed	Thu	Fri	Sat
Sales						

3. The bar chart of Fig. 21.8 gives the answer to a questionnaire, in which people were asked how often they had been to the cinema in the last month.
   a) How many people went twice or more?
   b) How many people were asked?

4. A group of 60 girls were asked which sport they were best at. The results are shown in the pie chart of Fig. 21.9.
   a) How many girls are best at netball?
   b) 5 girls are best at athletics. What is the angle for that sector of the pie?

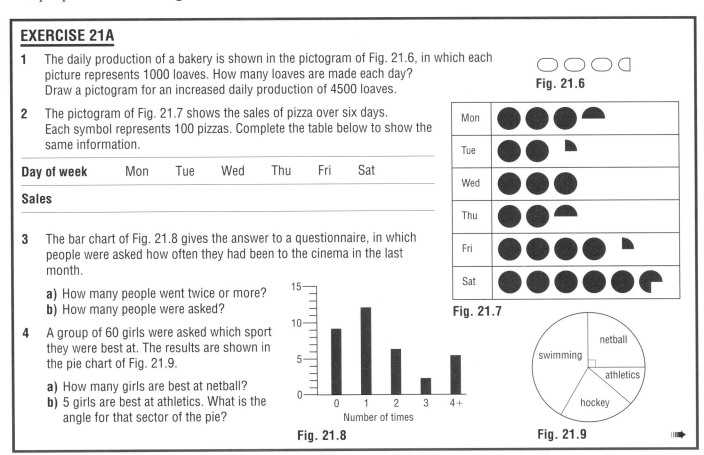

5. There are 40 teachers in a school. Their main subjects are shown in the pie chart of Fig. 21.10.

   a) There are 8 Maths teachers. What is the angle in that sector of the pie?
   b) The angle in the Languages sector is 99°. How many Languages teachers are there?

6. The table below shows where the cars owned by a group of people were manufactured. Construct a bar chart on Fig. 21.11 to show this information.

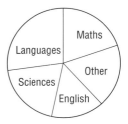

Fig. 21.10

Country of manufacture	Britain	France	Germany	Japan	Other
Frequency	8	9	12	6	15

Fig. 21.11

21.2 Construction of pie charts

When constructing a pie chart, divide each frequency by the total frequency, then multiply by 360°. This gives the angle of the sector.

■ Example 21.4

After tax and mortgage payments, a woman has £180 per week to spend. The table below shows how it is spent.

Food	Entertainment	Car	Other
£65	£27	£42	£46

Construct an accurate pie chart showing this information.

Solution

The total amount of money is £180. For each individual amount, divide by 180 and multiply by 360°.

Food: $\frac{65}{180} \times 360° = 130°$

Entertainment: $\frac{27}{180} \times 360° = 54°$

Car: $\frac{42}{180} \times 360° = 84°$

Other: $\frac{46}{180} \times 360° = 92°$

Note that the sum of these angles is 360°. Now construct the pie chart, dividing the circle into sectors corresponding to each group. The result is Fig. 21.12.

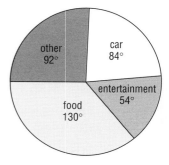

Fig. 21.12

PICTURES OF DATA

Key point

■ When constructing a pie chart, add up the angles to ensure that their sum is 360°.

EXERCISE 21B

1 A library conducts a survey into what sort of books are borrowed most frequently. The results for 400 borrowings are in the table below. Construct an accurate pie chart on Fig. 21.13 to show the information.

Fiction	Biography	Travel	Other
130	50	30	190

2 The 90 CDs in a collection are classified as shown below. Construct a pie chart to illustrate the figures.

Type of CD	Number
Classical	23
Rock	18
Jazz	38
Country	11

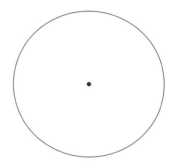

Fig. 21.13

3 A population survey of a country shows that 10% live in city centres, 30% in suburbs, 35% in small towns and 25% in the country. Show this information on a pie chart.

21.3 Bar charts from frequency tables; frequency polygons

A bar chart can be drawn to illustrate a frequency table. If the data relates to a continuous range of numbers, the bars should touch each other. A **frequency polygon** shows frequencies by points on a graph. It can be obtained by joining up the tops of bars in a bar chart.

■ Example 21.5

The incomes of the employees of a firm are given in the frequency table below.

Income (£1000s)	15–20	20–25	25–30	30–35
Frequency	29	37	21	15

Construct a bar chart to show these figures.

Solution

Put the income of the employees along the scale of the *x*-axis. Draw vertical bars for the frequencies. The result is shown in Fig. 21.14. Notice that the bars are touching.

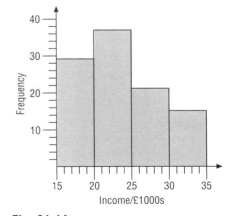

Fig. 21.14

PICTURES OF DATA

■ Example 21.6

A group of children were asked how many brothers or sisters they had. Five had none, twelve had one, four had two, and six had three or more. Illustrate this by a frequency polygon.

Solution

First draw a bar chart, using lines for the bars. Join up the tops of the bars as in Fig. 21.15.

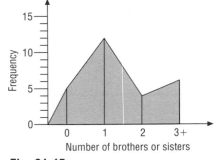

Fig. 21.15

Key point

■ If a bar chart displays continuous numerical information, the bars should touch each other. If the information is not continuous, the bars should not be touching.

EXERCISE 21C

1 The ages of the members of a squash club are given in the table below. Construct a bar chart on Fig. 21.16 to show this information.

Age	15–20	20–25	25–30	30–35
Frequency	18	29	14	9

2 The prices of the second-hand cars advertised in a paper were as given in the table below. Construct a bar chart to illustrate the information.

Price (£)	0–1000	1000–2000	2000–3000	3000–4000
Frequency	26	33	41	25

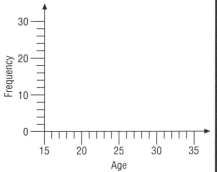

Fig. 21.16

3 The average daytime temperature was measured over a period. The results are below. Construct a bar chart to show the information.

Temperature (°C)	5–10	10–15	15–20	20–25	25–30
Frequency	23	27	34	19	8

4 The efficiency of the Royal Mail was tested by finding how long they took to deliver first class letters. The results are in the table below. Construct a frequency polygon on Fig. 21.17 to show the information.

Number of days	1	2	3	4
Frequency	16	10	5	2

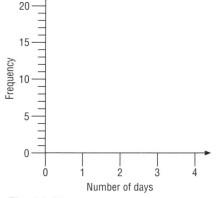

Fig. 21.17

21.4 Scatter diagrams

A **scatter diagram** shows whether there is a connection between two quantities. If the points lie roughly on a straight line, the quantities are **correlated**. If the line is sloping upwards, there is **positive correlation**. If the line is sloping downwards, there is **negative correlation**. If the points do not lie on a straight line at all, then there is **no correlation**.

Fig. 21.18 shows scatter diagrams for situations with positive correlation, negative correlation and no correlation.

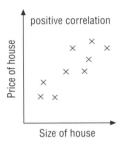 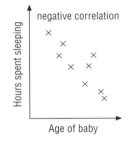 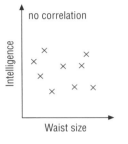 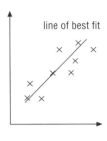

Fig. 21.18   Fig. 21.19

When points show correlation, the **line of best fit** is the straight line drawn through the points that is closest to them. See Fig. 21.19.

■ Example 21.7

On nine days of the year, Marianne measured the number of hours of daylight and the temperature at midday. Her results are in the table below. The length of the day is x hours, and the temperature is $y°C$.

Length of day (hours)	x	8	9	10	11	12	13	14	15	16
Temperature (°C)	y	7	6	6	10	15	12	21	25	14

Plot these points on a scatter diagram. What sort of correlation is there?
Draw a straight line through the points.
Predict the midday temperature for a day of length $12\frac{1}{2}$ hours.

Solution
The points are plotted in Fig. 21.20. Notice that the points lie roughly on a line going upwards.

There is positive correlation.
The line of best fit is drawn through the points.
The value of y corresponding to $x = 12\frac{1}{2}$ is 14.
The midday temperature might be 14°C.

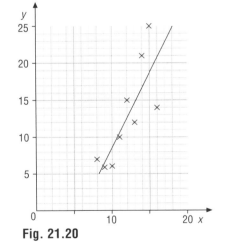

Fig. 21.20

Key point
■ A line of best fit through a scatter diagram goes through the 'middle' of the points. There should be about as many points above the line as below it. The line doesn't necessarily go through the first and last points, or through the origin.

© IT IS ILLEGAL TO PHOTOCOPY THIS PAGE

EXERCISE 21D

1 The table below gives the ages and prices of ten second-hand cars advertised in a newspaper. Plot them on Fig. 21.21. What sort of correlation is there?

Age (years)	5	3	8	2	2	5	7	8	6	4
Price (£)	2000	5000	1500	8000	6000	4500	2500	1000	2000	3000

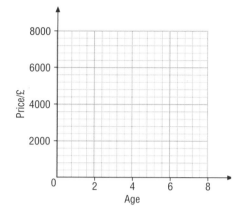

Fig. 21.21

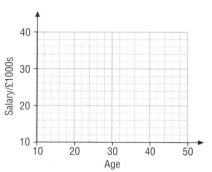

Fig. 21.22

2 A firm has nine employees. Their ages and salaries are given below. Plot them on Fig. 21.22. What sort of correlation is there?

Age (years)	25	34	18	45	30	50	26	22	38
Salary (£1000s)	19	27	15	25	22	34	28	20	30

3 On your graph for Question 1, draw a line of best fit through the points.
If a car is advertised at £4000, how old is it likely to be?

4 On your graph for Question 2, draw a line of best fit through the points. What might be the salary of someone aged 40?

5 In the sixth form of a school there are eight students taking both French and German. Their marks in mock exams are given below. Plot the points on a scatter diagram. Draw a line of best fit, and predict the mark for German of a student who got 65% in French.

French mark (%)	60	42	88	71	80	90	48	55
German mark (%)	51	37	52	84	82	82	42	50

Revision checklist

This chapter has revised:

21.1 Obtaining information from pictograms, bar charts and pie charts. Constructing pictograms and bar charts. ❏
21.2 Constructing pie charts. ❏
21.3 Bar charts of continuous numerical data. Frequency polygons. ❏
21.4 Scatter diagrams. Positive and negative correlation. Drawing a line of best fit, and using it to find values. ❏

CHAPTER 22 Analysing data

Chapter key points
- Read questions carefully to see which of the three averages, mean, median or mode, you are asked to find.
- Data such as 'numbers of people' consist of whole numbers. But the mean may be a fraction. The average number of passengers in a car might be 0.95 people per car. Don't round it to a whole number unless you are asked to.
- To find the median of an even number of values, say for eight values, take halfway between the fourth and fifth values. With an odd number of values, say nine values, take the fifth (that is the fifth *value*, not the number 5 itself).
- When finding the mean from a frequency table, divide by the total of the frequencies, not by the number of intervals.
- If a frequency table contains intervals of values, take the middle of each interval when estimating the mean.
- On a cumulative frequency curve, plot the points at the upper ends of the intervals (not the middles).

22.1 Averages

An **average** of a set of numbers tells us the approximate size of the numbers. There are different ways of measuring averages, which are used to give the best way of summarising data in different situations.

Mean If there are n numbers, add them and divide by n.
Median The middle number. Arrange the numbers in order, take the middle number (or the mean of the two middle numbers).
Mode The most frequent number.

To find the **range** of a set of numbers, subtract the lowest number from the highest number.

Note. When asked simply for the 'average', this implies the *mean*.

■ Example 22.1

The numbers below are the numbers of goals scored in 12 football matches. Find the mean and the mode of these numbers.

$$0 \quad 1 \quad 1 \quad 0 \quad 0 \quad 3 \quad 2 \quad 1 \quad 5 \quad 2 \quad 0 \quad 3$$

Solution
For the mean, find the sum of all the numbers.
$0 + 1 + 1 + 0 + 0 + 3 + 2 + 1 + 5 + 2 + 0 + 3 = 18$
Divide by 12, obtaining $1\frac{1}{2}$.
The mean number of goals is $1\frac{1}{2}$.
The most common number is 0, which occurs four times.
The mode is 0 goals.

ANALYSING DATA

■ Example 22.2

The weights, in kg, of nine children were recorded as below. Find the median weight and the range.

55 68 49 39 50 45 53 44 61

Solution

Rearrange the weights in increasing order:
39 44 45 49 50 53 55 61 68
The middle number is the fifth, 50.
The median weight is 50 kg.
For the range, subtract 39 from 68.
The range of the weights is 29 kg.
Note. There were an odd number of children. If there had been eight children, the median would be halfway between the fourth weight and the fifth weight.

Key points

- Read questions carefully to see which of the three averages, mean, median or mode, you are asked to find.
- Data such as numbers of people consists of whole numbers. But the mean may be a fraction. The average number of passengers in a car might be 0.95 people per car. Don't round it to a whole number unless you are asked to.
- To find the median of an even number of values, say eight values, take halfway between the fourth and the fifth values. With an odd number of values, say nine values, take the fifth (that is the fifth *value*, not the number 5 itself).

EXERCISE 22A

1 In ten innings, a cricket team makes the following scores. Find the mean score and the median score.

218 198 155 87 288 220 168 108 331 264

2 In a supermarket apples are sold in packs of six. The manageress orders an investigation into the number of bruised apples in the packs. The numbers of bruised apples in twelve packs are below. Find the mean number and the mode number.

0 1 2 1 0 0 1 1 5 1 3 2

3 The midday temperature at a resort is found over seven days. The results, in °C, are below. Find the median and the range of the temperatures.

27 30 23 25 34 28 31

4 Ten people estimate the number of cherries in a bowl. The results are below. Find the mean, median and range of the guesses.

95 106 172 70 100 127 130 200 150 166

5 The numbers below are the marks obtained by eleven children in an exam. Find the mean, median and range of these marks.

48 63 69 77 58 50 42 66 73 82 52

6 There are seven questions in an exam. Nine candidates take the exam, and the numbers of questions they attempt are shown below. Find the mean number and the mode number.

7 6 7 5 7 5 5 7 7

22.2 Averages from frequency tables

Averages can be found for data from a frequency table. The results may be approximate. The group with the greatest frequency is the **modal group**.

■ Example 22.3

Laura stands by the side of a road and makes a record of the number of passengers in the cars that pass. The results for 80 cars are given in the following frequency table:

Number of passengers	0	1	2	3	4
Frequency	35	26	10	6	3

Find the mean number of passengers per car and the median number.

Solution
Find the total number of passengers. Multiply each number of passengers by its frequency, then add.
$0 \times 35 + 1 \times 26 + 2 \times 10 + 3 \times 6 + 4 \times 3 = 76$
There are 80 cars in all. Divide 76 by 80, obtaining 0.95.
The *mean* number of passengers is 0.95.
Suppose the numbers were arranged in increasing order, as in Fig. 22.1. There would be 35 0s, then 26 1s, then 10 2s, and so on. Both the 40th and the 41st numbers would be 1.

000..........111111......222222.......333333444

Fig. 22.1

The *median* number of passengers is 1.

■ Example 22.4

The annual rainfall at a town, in inches, has been recorded for 50 years. The results are below. Find the modal class and estimate the mean annual rainfall.

Rainfall (in)	20–30	30–40	40–50	50–60
Frequency	13	19	12	6

Solution
The modal class is the class with the largest frequency.
The modal class is 30–40 inches.
We do not know exact figures for the rainfall. Assume that the numbers are evenly spread within each interval, i.e. that the rainfall in the 20–30 interval averages at 25 inches per year. Make a similar assumption for the other intervals. Rewrite the table:

Rainfall (in)	25	35	45	55
Frequency	13	19	12	6

Now work out the total rainfall:
$25 \times 13 + 35 \times 19 + 45 \times 12 + 55 \times 6 = 1860$ inches
The data covers 50 years. Divide 1860 by 50, obtaining 37.2.
The mean annual rainfall is approximately 37 inches.

© IT IS ILLEGAL TO PHOTOCOPY THIS PAGE

ANALYSING DATA

> **Key points**
> - When finding the mean from a frequency table, divide by the total of the frequencies, not by the number of intervals.
> - If a frequency table contains intervals of values, take the middle of each interval when estimating the mean.

EXERCISE 22B

1 A group of 50 children were asked how many foreign countries they had visited. The results are in the table below. Find the mean and median number of countries visited.

Number of countries	0	1	2	3	4	5	6	7
Frequency	10	12	9	8	4	5	1	1

2 A school has 80 teachers. The head teacher finds how many days the teachers have taken off for sickness in the previous term. The results are below. Find the mode, mean and median number of days.

Number of days	0	1	2	3	4	5	6	7	8	9
Frequency	21	13	16	8	7	8	2	4	0	1

3 Over 60 days, the midday temperature at a resort was measured. The results, in °F, are below. Find the modal class and estimate the mean temperature.

Temperature (°F)	55–60	60–65	65–70	70–75	75–80
Frequency	9	10	17	15	9

4 After returning from holiday, 100 people were asked how much they had spent. The results are below. Find the modal class and estimate the mean amount.

Amount spent (£)	50–100	100–150	150–200	200–250	250–300
Frequency	18	29	31	18	4

22.3 Cumulative frequency

Suppose you have a set of data. The **cumulative frequency** at value x is the number of data less than x. A **cumulative frequency curve** or **ogive** is a graph of the cumulative frequency.

- At the median, the cumulative frequency is $\frac{1}{2}$ of the total frequency.
- At the **lower quartile**, the cumulative frequency is $\frac{1}{4}$ of the total frequency.
- At the **upper quartile**, the cumulative frequency is $\frac{3}{4}$ of the total frequency.
- The **inter-quartile range** is the difference in values between the quartiles. It tells us how widely spread the data is.

Example 22.5

The cumulative frequency graph of Fig. 22.2 shows the prices of 200 houses in a town in 1995.

a) How many houses cost more than £85 000?
b) Find the median and the inter-quartile range of the prices.

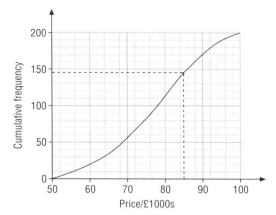

Fig. 22.2

Solution
a) Follow the dotted line from a price of £85 000 to a cumulative frequency of 145.
Hence 145 houses cost less than £85 000.
Subtract from 200, to obtain 55.
55 houses cost more than £85 000.

b) The median corresponds to a cumulative frequency of 100.
From the graph, this is reached at a price of £78 000.
The median is £78 000.
The quartiles correspond to cumulative frequencies of 50 and 150. From the graph, the quartile values are £69 000 and £86 000. The difference is £17 000.
The inter-quartile range is £17 000.

Example 22.6

The graph in Example 22.5 refers to 1995. A frequency table for 1990 is below.

Price range (£1000s)	50–60	60–70	70–80	80–90	90–100
Frequency	35	71	54	24	16

a) Find the cumulative frequencies and plot a cumulative frequency graph on the same axes as in Example 22.5.
b) A house cost £67 000 in 1990. What would it cost in 1995?

Solution
a) Add a third row to the table, adding up the frequencies as you go along:

Price range (£1000s)	50–60	60–70	70–80	80–90	90–100
Frequency	35	71	54	24	16
Cumulative frequency	35	106	160	184	200

The points on the graph are placed at the *end-points* of the intervals: 35 houses cost £60 000 or less, so the point (60, 35) lies on the graph. The dotted curve in Fig. 22.3 shows the graph for 1990.

b) The *order* of the prices will not have changed much between 1990 and 1995. On the dotted graph (1990), a price of £67 000 is at a cumulative frequency of 85; on the filled in graph (1995) a cumulative frequency of 85 is at a price of £75 000.
In 1995, the house would cost £75 000.

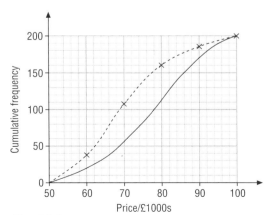

Fig. 22.3

ANALYSING DATA

Key point

■ On a cumulative frequency curve, plot the points at the upper ends of the intervals (not the middles).

EXERCISE 22C

1 The cumulative frequency graph of Fig. 22.4 shows the marks obtained in an exam by 1000 candidates. If the pass mark was 50%, how many passed? Find the median mark and the inter-quartile range.

2 A man did a survey of the salaries of jobs advertised in a newspaper. The results are in the table below.

Salary (£1000s)	12–14	14–16	16–18	18–20	20–22	22–24
Frequency	5	17	25	30	13	10

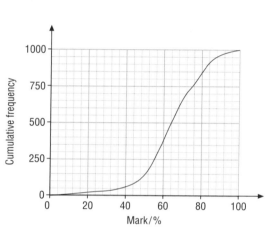

Fig. 22.4

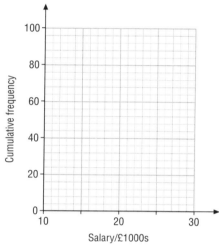

Fig. 22.5

a) Complete a row of the cumulative frequencies.
b) Draw a cumulative frequency curve on Fig. 22.5.
c) Find the median, the quartiles, and the inter-quartile range.
d) What proportion of jobs pay more than £21 000?

3 80 dogs were weighed. The table below gives the results.

Weight (kg)	10–20	20–30	30–40	40–50
Frequency	13	35	26	6

a) Find the cumulative frequencies.
b) Draw a cumulative frequency curve.
c) From your curve find the median, the quartiles and the inter-quartile range.

4 Over 60 days of the summer, the midday temperature was found in two tourist resorts, A and B. The table below gives the results.

Temperature (°C)	60–65	65–70	70–75	75–80	80–85	85–90
Resort A frequency	2	8	18	19	10	3
Resort B frequency	6	12	13	12	9	8

a) Find the cumulative frequencies for both the resorts, and draw the cumulative frequency curves on Fig. 22.6. What is the difference between the curves?
b) Find the inter-quartile ranges for the two resorts. What does the difference between the ranges tell you about the two resorts?

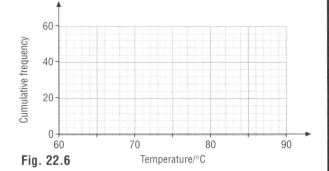

Fig. 22.6

5 Fig. 22.4 of Question 1 refers to the first of two exams. The table below gives the marks for the second exam.

Mark	0–19	20–39	40–59	60–79	80–100
Frequency	121	301	328	180	70

a) Find the cumulative frequencies and plot a cumulative frequency graph on Fig. 22.4 above.
b) Amy scored 68 in the first exam. What do you think she scored in the second?

Revision checklist

This chapter has revised:

22.1 Finding averages of a set of data – the mean, median and mode. The range of a set of data. ❏
22.2 Finding averages from frequency tables. ❏
22.3 Plotting a cumulative frequency curve, and using it to find the median, the quartiles and the inter-quartile range. ❏

CHAPTER 23 *Probability*

Chapter key points
- Suppose an experiment has n equally likely results. The probability of each individual outcome is $\dfrac{1}{n}$.
- If one pupil is picked at random from a class of ten girls and eight boys, the probability that a girl is picked is not $\frac{1}{2}$ – a girl is more likely to be picked than a boy because there are more girls in the class.
- The first branch of a tree diagram gives the outcomes of the first experiment, and the second branches give the outcomes of the second. Don't put the outcomes of both experiments on the first branch.

The **probability** of an event tells us how likely it is. Probabilities can be shown on a number line between 0 and 1, as shown in Fig. 23.1. An event which is certain has probability 1. An event which is impossible has probability 0.

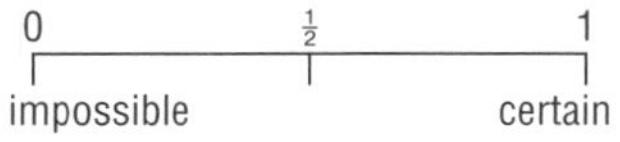

```
0           ½            1
|           |            |
impossible              certain
```
Fig. 23.1

The probability of an event A is written P(A). Suppose that P(A) = p, i.e. the probability that A happens is p. The probability that A does *not* happen is $1 - p$.

23.1 Single probability

On a fair cubical die there are six faces, each of which is equally likely to come uppermost. Hence the probability of each face coming uppermost is $\frac{1}{6}$. In general, if an experiment has n equally likely results, then the probability of each result is $\dfrac{1}{n}$.

■ Example 23.1

In a standard pack of cards there are four suits – Spades, Hearts, Diamonds and Clubs. Each suit contains 13 cards – Ace, King, Queen, Jack, 10 down to 2. A card is drawn from a well shuffled pack. What is the probability that it is a Heart?

Solution
Each of the four suits is equally likely to be drawn. Hence the probability of each suit is $\frac{1}{4}$.
The probability of a Heart is $\frac{1}{4}$.
Note. Another way of reaching the same answer is this. There are 13 Hearts, out of a total of 52 cards. Hence the probability of a Heart is $\frac{13}{52}$, i.e. $\frac{1}{4}$.

■ Example 23.2

A weather forecaster reckons that the probability of some rain tomorrow is $\frac{2}{5}$. What is the probability that it will not rain tomorrow?

Solution
The probability of rain is $\frac{2}{5}$. To find the probability of no rain, subtract from 1.
The probability of no rain is $\frac{3}{5}$.

PROBABILITY

Key points

- Suppose an experiment has n equally likely results. The probability of each individual outcome is $\frac{1}{n}$. Suppose Anne is in a class with ten girls and eight boys. If one pupil is picked at random, the probability that Anne is picked is $\frac{1}{18}$.
- The probability that a *girl* is picked from Anne's class is not $\frac{1}{2}$ – a girl is more likely to be picked than a boy because there are more girls in the class.

EXERCISE 23A

1. Fig. 23.2 shows a number line. On it place the probabilities of the events:

   a) that the day after next Sunday will be a Monday
   b) that it will rain next Sunday
   c) that the day after next Sunday will be a Tuesday.

   Fig. 23.2

2. Estimate the probabilities of the following:

   a) that a team from Europe will win the next World Cup for soccer
   b) that there will be a manned landing on Mars by 2020
   c) that you will study A-level mathematics next year.

3. A coin has two sides, Heads and Tails. The coin is fair. If it was spun, what is the probability that it would come up Heads?

4. A fair cubical die is rolled. What is the probability that the uppermost face is 6?

5. A fair octahedral (eight-sided) die has the numbers 1 to 8 on its faces. It is rolled. What is the probability that it lands on the face labelled 5?

6. In my pocket there are seven 10p coins and five 2p coins. I draw out a coin at random. What is the probability that it is a 10p coin?

7. Fig. 23.3 shows a box containing 7 buttons, some white and some coloured. If one is picked at random, what is the probability that it is white?

   Fig. 23.3

8. A box contains 50 sweets, of which one is a toffee.

   a) Joseph picks a sweet at random. What is the probability that it is not a toffee?
   b) Kay picks some sweets at random. The probability that she picks the toffee is $\frac{1}{10}$. How many sweets did she pick?

9. A pub raffle has 100 tickets, only one of which wins a prize.

   a) Stan buys 8 tickets. What is the probability that he will win a prize?
   b) Olive buys some tickets. The probability that she will win the prize is 0.3. How many tickets did she buy?

10. There are six letters in the word MURRAY. Each is written on a piece of paper, which are then put into a hat. One piece is drawn out. What is the probability that the letter shown is:

    a) Y b) R c) T?

11. A fairground game involves a wheel divided into ten equal sectors. One sector is black, four are coloured and five are white. A pointer is spun until it comes to rest over one of the sectors. Find the probabilities that it points at:

    a) the black sector b) a coloured sector

Fig. 23.4

12. A card is drawn from a standard pack. What is the probability that it is:

    a) a Queen   b) the Queen of Spades   c) not a Queen?

13. The probability that it will snow some time on Christmas day is $\frac{2}{7}$. What is the probability that it will not snow on Christmas day?

14. The probability that an entry in the National Lottery wins a prize is about $\frac{1}{70}$. What is the probability that an entry does not win a prize?

23.2 Listing the possibilities

■ Example 23.3

Fig. 23.5 shows two fair spinners. If both are spun, one possible outcome is A for the first spinner and 1 for the second, as shown in the diagram. Write this as A1. List the possible outcomes, and find the probability that the outcome is C3.

Fig. 23.5

Solution

The first spinner could be A, B or C, and the second spinner could be 1, 2 or 3.
The outcomes are A1, A2, A3, B1, B2, B3, C1, C2 and C3.
There are 9 equally likely outcomes.
The probability of C3 is $\frac{1}{9}$.

EXERCISE 23B

1. In a restaurant, the first course is either Avocado or Borscht. The second course is Casserole, Daube or Escalopes. List the possible choices for a two-course meal. If a diner picks her meal at random, what is the probability that she chooses AC (Avocado and Casserole)?

2. Dave has T-shirts in red, blue and pink. He has trousers in brown, green and white. List the possible combinations he could wear. If he picks the clothes at random, what is the probability that he wears RB (red T-shirt, brown trousers)?

3. Two fair tetrahedral (four-sided) dice each have the numbers 1 to 4 on their faces. They are thrown together, and the total of the numbers they land on is found. The table below gives the scores:

		Score on first die			
		1	2	3	4
Score on second die	1	2	3	4	5
	2	3	4	5	6
	3	4	5	6	7
	4	5	6	7	8

Find the probabilities that:

a) the total is 8   b) the total is 5   c) the total is 9

4 Two fair six-sided dice are thrown. The table below gives the scores:

		Score on first die					
		1	2	3	4	5	6
Score on second die	1	2					7
	2	3	4				
	3		5			8	
	4	5				9	
	5						
	6						12

    a) Complete the table.
    b) What are the probabilities that the total is
        i) 2   **ii)** 6?
    c) What is the most likely score? What is its probability?

5 Two fair coins are spun. List the possible outcomes. What is the probability that HH (double Head) is achieved?

6 Three runners A, B and C, run a race. List the possible orders in which they could finish. If all outcomes are equally likely, find the probability that they finish in the order ABC.

23.3 Combinations of probabilities

If two events do not affect each other, they are **independent**. If two independent events A and B have probabilities p and q respectively, then the probability of A and B happening is $p \times q$.

If events cannot happen together, then the probability of one of them occurring is found by *adding* their probabilities. If A and B have probabilities p and q respectively, the probability of *either* A *or* B happening is $p + q$.

A **tree diagram** is used to show the probabilities of successive events.

■ Example 23.4

A fair die is rolled and a fair coin is spun. What is the probability that the die gives a 6 and the coin gives a Head?

Solution
For the die, the probability of a six is $\frac{1}{6}$. For the coin, the probability of a Head is $\frac{1}{2}$. These events do not affect each other, hence they are independent. Multiply $\frac{1}{6}$ by $\frac{1}{2}$, obtaining $\frac{1}{12}$.
The probability is $\frac{1}{12}$.

■ Example 23.5

A menu contains 6 choices for the first course, of which 2 are vegetarian. It contains 8 choices for the second course, of which 3 are vegetarian. A diner picks both courses at random. Complete the tree diagram of Fig. 23.6. What is the probability that both dishes are vegetarian?

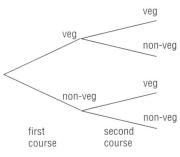

Fig. 23.6

Solution

The probability of a vegetarian choice for the first course is $\frac{2}{6}$, i.e. $\frac{1}{3}$. The probability of a non-vegetarian choice is $\frac{2}{3}$. Put these probabilities on the first branch of the tree.

The probabilities for the second course are $\frac{3}{8}$ and $\frac{5}{8}$. Put these on the second branches. Fig. 23.7 shows the result.

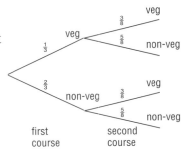

Fig. 23.7

The top branch corresponds to a vegetarian choice for both courses. Multiply the probabilities along this branch. The product of $\frac{1}{3}$ and $\frac{3}{8}$ is $\frac{1}{8}$.
The probability is $\frac{1}{8}$.

■ Example 23.6

What is the probability that the diner of Example 23.5 has exactly one vegetarian dish?

Solution

Use the tree diagram above.
P(vegetarian first course then non-vegetarian second course) = $\frac{1}{3} \times \frac{5}{8} = \frac{5}{24}$
P(non-vegetarian first course then vegetarian second course) = $\frac{2}{3} \times \frac{3}{8} = \frac{6}{24}$
Add these, obtaining $\frac{11}{24}$.
The probability of exactly one vegetarian course is $\frac{11}{24}$.

Key point

■ The first branches of a tree diagram give the outcomes of the first experiment, and the second branches give the outcomes of the second. Don't put the outcomes of both experiments on the first branch!

EXERCISE 23C

1. A card is drawn from a standard pack, and immediately afterwards a fair die is rolled. What is the probability that the card is an Ace and the die gives a 6?

2. A fair die is rolled twice. What is the probability that the first roll gives an even number and the second roll gives an odd number?

3. In a multiple choice exam, the first question has five possible answers, and the second question has four possible answers. A candidate answers both questions at random. What are the probabilities that:

   a) both questions are right
   b) both questions are wrong?

4. A biased coin is such that the probability of Heads is $\frac{1}{3}$. The coin is spun twice. Find the probabilities of:

   a) two heads
   b) two tails

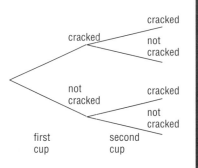

Fig. 23.8

5. A machine makes tea cups. One in twenty of the cups is cracked. Two cups are picked at random. Show probabilities on each branch of the tree diagram of Fig. 23.8.

   a) What is the probability that both cups are cracked?
   b) What is the probability that neither cup is cracked?

6. One bag contains 3 white and 4 black balls. A second bag contains 2 white and 3 black balls.
   I pick one ball from each bag. Complete the tree diagram of Fig. 23.9.

   a) What is the probability that both balls are white?
   b) What is the probability that one ball is black and one is white?

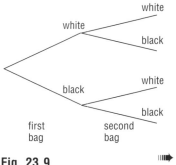

Fig. 23.9

7   Question 11 in Exercise 23A involved a spinner at a fairground.
    I have two goes on the spinner. Complete the tree diagram of Fig. 23.10.

    a) What is the probability that I get black both times?
    b) What is the probability that I get black exactly once?

8   A travelling salesman reckons that one in eight of his calls results
    in a sale. He makes two calls. Draw a tree diagram showing
    the results of the calls.

    a) What is the probability that both calls result in a sale?
    b) What is the probability that from the two calls, he makes
       exactly one sale?

9   Fig. 23.11 shows a map of part of a maze. Algernon the rat
    enters the maze at A.
    At each corner he chooses each of the available routes
    with equal probability.
    Find the probabilities that:

    a) he goes to B
    b) he goes to C

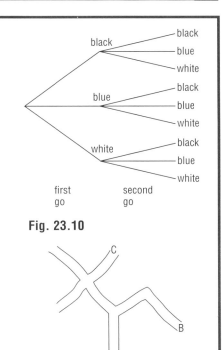

Fig. 23.10

Fig. 23.11

Revision checklist

This chapter has revised:

23.1 The probability of one of n equally likely outcomes. ❏
23.2 Listing the possible outcomes of more complicated experiments. ❏
23.3 Combinations of events, and the use of a tree diagram to find
probabilities connected with two successive events. ❏

Mixed exercise 4

1. A crate of beer contains 14 bottles of light ale and 10 bottles of brown ale. A bottle is picked at random. What is the probability that the bottle contains brown ale? Give your answer as a fraction in its simplest form.

2. Susie wants to find out how often people go to the dentist. Devise a questionnaire she could use.

3. Fig. M4.1 is a scatter graph for the prices of ten London houses and their distances from the centre of London.

   a) What sort of correlation does the graph show?
   b) Draw a line of best fit through the points.
   c) From your line, predict the price of a house that is 12 miles from the centre.

4. Fig. M4.2 is a pie chart which shows how the votes in a Scottish by-election were divided between four parties.

   a) The Labour Party received 18 000 votes, and the angle of the Labour slice of the pie is 100°. How many voters were there in total?
   b) The Lib Dem Party received 10 800 votes. What is the angle in their slice of the pie?

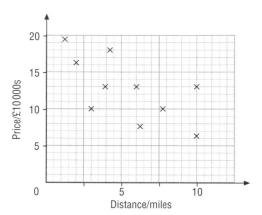

Fig. M4.1

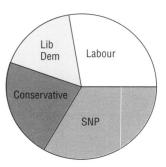

Fig. M4.2

5. The pictogram of Fig. M4.3 shows the sales of cars in a showroom over three quarterly periods. Each symbol represents ten cars.

Fig. M4.3

   a) How many cars were sold in each of the quarters?
   b) In the fourth quarter of the year 45 cars were sold. Extend the pictogram to include the fourth quarter.

6. Every morning for ten days the temperature (in °F) of the water in a swimming pool is found. The results are below. Find the mean temperature and the range of temperatures.

   58° 59° 60° 60° 63° 64° 63° 63° 60° 58°

7. Denise needs a replacement for a watch battery. In a drawer there are 20 similar-looking batteries, of which 7 will work for her watch.

   a) If she picks out a battery at random, what is the probability it will work in her watch?

   Nigel needs a battery for his calculator. If he picks a battery at random from the drawer, the probability that it will work in his calculator is $\frac{1}{4}$.

   b) How many batteries in the drawer will work in his calculator?

8. Twelve electrical components are kept switched on until they fail. The times to failure, in hours, are given below.

   23 12 39 17 5 24 18 31 40 22 19 11

   Find the median time to failure.

9. Jane has three cards, labelled A, B and C. The cards are shuffled and the top one dealt out.

   a) What is the probability that the A card is dealt?
   The card is returned, and the cards are shuffled again. The top card is dealt out.
   b) One possible result for the two dealings is BC (first card B, second card C). List all the possible results.
   c) What is the probability that the A card is dealt out twice?

MIXED EXERCISE 4

10 There are four local newspapers in a town. A survey was undertaken into which paper people read most often. The results are below.

Mail Post Post Mail News Bugler Mail Bugler Post Post
News Bugler Mail News News Bugler Mail News Post
Bugler Bugler Post News Mail News Post Bugler Mail
News News Post Post Post Bugler Mail Mail

a) Complete the tally chart below. Find the frequencies.

News

Mail

Post

Bugler

b) Draw a bar chart on Fig. M4.4 to show the information.

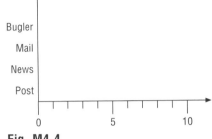

Fig. M4.4

The following questions are suitable for Intermediate level.

11 The weight, w grams, of 80 portions of fish are found. The values are given in the table below.

Weight (grams)	$90 \leq w < 100$	$100 \leq w < 110$	$110 \leq w < 120$	$120 \leq w < 130$
Frequency	13	25	31	11

a) Estimate the mean weight of a portion of fish.
b) On Fig. M4.5 construct a bar chart to illustrate the data.
c) Which interval contains the median weight?

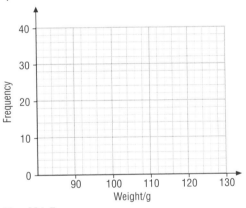

Fig. M4.5

Fig. M4.6

12 At a village fête, there is a Lucky Dip stall and a 'Pick a Number' game stall. In the Lucky Dip, $\frac{3}{4}$ of the parcels contain a prize. In the game, $\frac{1}{8}$ of the numbers will win a prize. Jake has a go at both the stalls.

a) Complete the tree diagram of Fig. M4.6.
b) What is the probability that he will win a prize on both stalls?
c) What is the probability he will win at least one prize?

13 Fig. M4.7 is a cumulative frequency diagram for the amounts spent by 100 shoppers in a supermarket one Saturday morning.

a) What is the median amount spent?
b) What is the inter-quartile range of the amount spent?

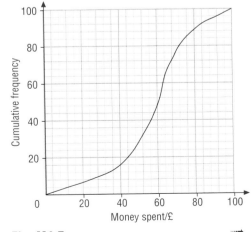

Fig. M4.7

On the following Saturday, a similar survey was done. The table below gives the amounts spent.

Amount spent (£)	0–20	20–40	40–60	60–80	80–100
Frequency	14	24	21	19	22

    c) Find the cumulative frequencies.
    d) Plot a cumulative frequency graph on Fig. M4.7.
    e) Mrs Evans went shopping on both Saturdays. She spent £60 on the first Saturday. What do you think she spent on the second Saturday?

14 Anne spins a bent coin 100 times, and obtains 40 Heads.

    a) What is the approximate probability that this coin will give Heads?
    Brian spins the same coin 80 times.
    b) How many Heads do you think he obtained?

15 A tetrahedral (four-sided) die has the numbers 1, 2, 3 and 4 on its sides. When it is tossed, the score is the number on the face it lands on. The die is tossed twice, and the total score is found.

    a) What is the probability that the total score is 2?
    b) What is the probability that the total score is 6?

16 A Republican organisation undertakes a survey into British attitudes to the monarchy. Its questionnaire includes the following:

    Should certain people hold power just because of their birth?

Is this question fair? Write an amended question.

MOCK EXAMINATIONS

Mock exam 1 (Foundation)

Marks

1. Write 476 in words. [2]

2. Shade a third of the shape of Fig. E1.1. [2]

Fig. E1.1

3. a) The train journey between Alphaville and Betatown takes 11 minutes. If it leaves Alphaville at 0947, when does it arrive at Betatown? [1]
   b) The train reaches Gammacity at 1009. How long did the train take between Betatown and Gammacity? [2]

4. Lionel's recipe for pastry uses butter and flour in the ratio 2 : 3. How much flour is needed to go with 10 ounces of butter? [2]

5. From the list of numbers 4, 5, 6, 8, write down:
   a) a prime number
   b) a square number [2]

6. a) A field is a square with side 30 m. What is the area of the field in m^2? [1]
   b) A sugar cube has side 1.2 cm. What is the volume of the cube in cm^3? Give your answer correct to 1 decimal place. [2]

Ravioli	£6.35
Lasagne (ringed)	£7.95
Fettuccine	£5.50
Osso buco (ringed)	£6.85

Fig. E1.2

7. 'Bella Nosh' is a meal delivery service. Part of its menu is shown in Fig. E1.2. What is the total cost of the ringed items? [2]
   The delivery charge is 10% of the cost of the food. What is the total to be paid? [2]

8. Fig. E1.3 shows a tessellation of triangles. Show how to continue this tessellation with three more triangles. [3]

Fig. E1.3

   These triangles have equal sides. What is the mathematical name for this sort of triangle? [1]
   These triangles have equal angles. What is each angle? [2]

9. The map of Fig. E1.4 shows two French towns, Tours and Le Mans. The map has the scale 1 cm to 20 km.
   a) What is the real distance between the towns? [2]
   b) Measure the bearing of Tours from Le Mans. [2]

Fig. E1.4

10. A bag contains 12 marbles, of which 5 are red, 3 brown and the rest blue. A marble is drawn out.
    a) What is the probability that the marble is brown? [1]
    b) What is the probability that the marble is not blue? [2]
    c) What is the probability that the marble is white? [2]

11. a) The first four odd numbers are 1, 3, 5 and 7. Write down the next two odd numbers. [1]
    b) The sum of the first four odd numbers is 16. Find the sum of the first five odd numbers. [1]
    c) Complete the table.

Sum of first odd number	1
Sum of first two odd numbers	
Sum of first three odd numbers	
Sum of first four odd numbers	16
Sum of first five odd numbers	

[3]

d) What sort of numbers appear in the right-hand column of the table? [2]

12 The table below gives the midday temperatures (in °C) during a cold spell.

Mon	Tue	Wed	Thu	Fri
−3°	−4°	1°	4°	4°

a) What was the lowest temperature? [1]
b) What was the rise in temperature between Monday and Friday? [1]
c) What was the greatest rise in temperature over one day? [2]

13 The shape S of Fig. E1.5 is reflected in the dotted line. Draw the reflected shape. [3]

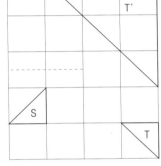

The shape T has been enlarged to T'. What is the scale factor of the enlargement? [2]

Fig. E1.5

14 I can go from London to Dover by train or by coach. I can cross the Channel from Dover to Calais by train, ferry or hovercraft. One possible way from London to Calais is Train–Ferry (TF). List all the other ways. [2]

I am offered a lift by car to Dover. How many ways are there now of getting from London to Calais? [2]

15 The dots of Fig. E1.6 are 1 cm apart. On the dots draw the net for a cuboid which is 3 cm by 2 cm by 1 cm. One face has been drawn for you. [3]

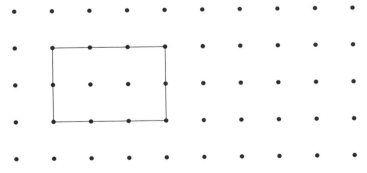

Fig. E1.6

16 The 'Burgerbaron' sells four types of burger, A, B, C and D. The sales during a four-hour period were recorded as below.

```
C D B B A B B C D A B B C D C C
C A A B D C D B A A B C C D A B A
B A C
```

a) Record this information on the tally sheet below.

A
B
C
D [4]

b) The sales can be recorded on a bar chart. Complete the bar chart on Fig. E1.7. [3]

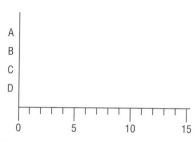
Fig. E1.7

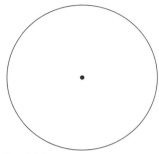
Fig. E1.8

c) The sales can also be recorded on a pie chart. Draw an accurate pie chart on Fig. E1.8, indicating the angles of each slice. [4]

d) The next four-hour period was busier. Twice as many of each type of burger were sold. What difference would this make to the pie chart? [2]

17 The conversion chart of Fig. E1.9 converts between £ and Swiss francs (SF).

a) i) What is £1.70 worth in SF? [2]
ii) What is 23SF worth in £? [2]

b) The exchange rate changes, so that £20 is now worth 52SF. Draw a line on the conversion chart for the new exchange rate. [4]

18 The equation $v = u + at$ occurs in Physics.

a) Find v if $u = 30$, $a = 10$ and $t = 4.5$. [2]
b) Suppose that $v = 50$, $u = 30$ and $a = 10$. Find t. [3]

19 Janet bought 49 cans of fizzy drink at 42p each.

a) *Without using a calculator*, find the total cost of the cans. Show all your working. [3]
b) Janet is not very confident about her arithmetic. Show how Janet could find a rough estimate to confirm that her answer is correct. [2]

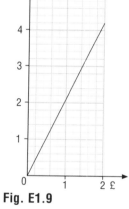
Fig. E1.9

20 An independent college receives its income partly from the Government, partly from fees and partly from investments. The Government provides $\frac{1}{8}$ of the income and the fees provide $\frac{3}{5}$.

a) What part of the income comes from investment? Give your answer as a fraction. [2]
b) The total income per term is £600 000. What is the income from fees? [2]

21 Fig. E1.10 shows a diagram of a motorway. It is not to scale. It gives the distances in miles between junctions. The junctions are either exits, with numbers, or service areas, marked S. The distances from two exits to two towns are also given.

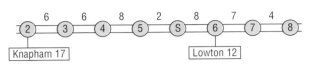

Fig. E1.10

a) How far is it from Junction 2 to the service area shown? [2]
b) A motorist drives from Knapham to the motorway at Junction 2, then along the motorway to Junction 6, then to Lowton. How far has the motorist driven? [2]
c) Between Junction 2 and Junction 6, the motorist drove at a speed of 60 m.p.h. How long did this part of the journey take? [2]

22 On the grid of Fig. E1.11, mark the points (1, 3) and (4, 0). Join them with a straight line. [2]

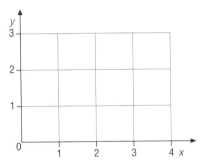

Fig. E1.11

The point $(x, 1)$ lies on this line. What is x? [1]

[Total 100]

Mock exam 2 (Foundation)

Marks

1. Write in figures: two thousand and thirty-six. [1]

2. Felix goes shopping, and buys the items shown in Fig. E2.1. Find the total bill. [2]

```
6 litres mineral water    £5.42
12 packets crisps         £2.75
3 lb oranges              £2.68
1 box of chocolates       £2.84
```

Fig. E2.1

How much change would Felix get from a £20 note? [1]

3. a) During a sale, all prices are reduced by a quarter. What is the reduced price of a tie that originally cost £8.00? [1]
   b) A legacy is divided equally between three brothers. Each brother then has to pay 20% of his legacy in inheritance tax. What fraction of the original legacy does each brother receive? [2]

4. Sid buys a jacket at £43.00 and two shirts at £15.00 each. He is given a discount of 10%. How much does he have to pay? [3]

5. Fig. E2.2 shows a sequence of patterns of dots.

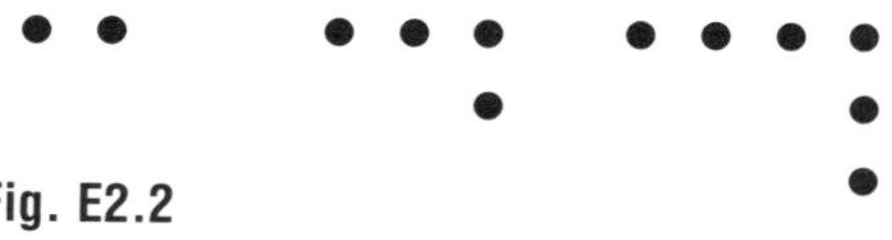

Fig. E2.2

   a) Draw the next pattern. [2]
   b) How many dots are there in the sixth pattern? [2]
   c) Suppose there are x dots in one of the patterns. How many dots are there in the next pattern? [2]

6. Annabel is 5 feet and 6 inches high.

   a) What is her height in inches? (1 foot = 12 inches) [1]
   b) What is her height in centimetres? (1 in = 2.54 cm) [2]
   c) What is her height in metres? [1]

7. Fig. E2.3 shows three scatter diagrams. Which of them could represent the relation between the following quantities?

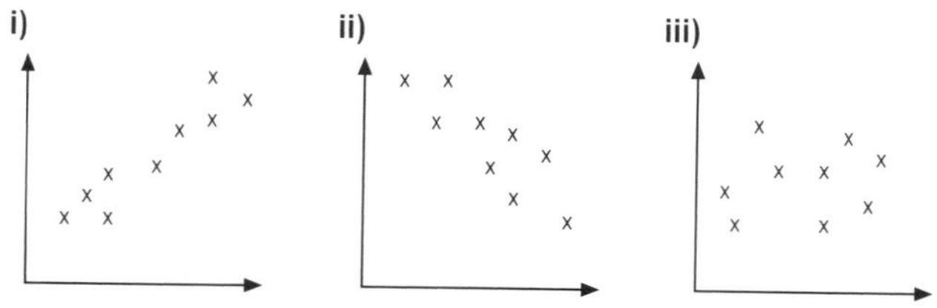

   i)   ii)   iii)

Fig. E2.3

   a) The amount owing on a mortgage and the number of years it has run. [1]
   b) The average annual income of a country and its average temperature. [1]
   c) The acreage of a farm and its price at auction. [1]

© IT IS ILLEGAL TO PHOTOCOPY THIS PAGE

8 A running track is in the shape of a circle with radius 60 m.

  a) What is the length of the running track, to the nearest m? [2]
  b) What area is enclosed within the track, to the nearest 100 m^2? [2]
  c) For his training, Joe has to run a total of 10 000 m. How many times will he have to run round the track? Give your answer correct to the nearest whole number. [2]

9 Fig. E2.4 shows two Chinese characters.
How many lines of symmetry are there in character **a)**? [1]
What is the order of rotational symmetry of character **b)**? [1]

Fig. E2.4

10 Rahim is required to find the mean of the ten numbers:

   12 13 15 12 16 17 13 16 17 19

  a) He obtains the answer 23. Explain why his answer must be wrong. [2]
  b) Find the median of the numbers. [2]

11 Complete the table below for the formula $y = 2x - 1$.

x	1	2	3	4	5
y					

[3]

Plot the graph of the function on Fig. E2.5. [3]
Use your graph to find the value of x for which $6 = 2x - 1$. [2]

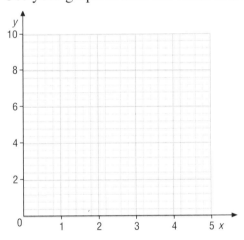

Fig. E2.5

12 Fig. E2.6 shows some of the programmes on one television channel during an evening.

6.00	News
6.25	Weather
6.35	*The Environment.* Investigation into polystyrene pollution
7.00	Comedy. *A Cut Off The Joint*
7.40	Sports review

Fig. E2.6

  a) How long does the News last? [1]
  b) Anita switches on at 6.30, and switches off 40 minutes later. Which programmes has she seen at least part of? [2]
  c) Anita did not see all of the comedy. What percentage of the comedy did she miss? [2]

13 Fig. E2.7 is a pictogram in which each symbol represents 100 CDs. It shows the numbers of CDs sold in a music shop. Write down how many CDs were sold in each of the three categories. [2]

Rock	◯ ◯ ◯ ◯ ◯
Jazz	◯ ◯ ◖
Classical	◯ ◯ ◯ ◔

Fig. E2.7

A mystery prize is given to the buyer of one of these CDs. What is the probability that the prize is won by someone who buys a jazz CD? [2]

There were 125 folk music CDs sold. Add another row to the pictogram to illustrate this. [3]

14 Sean and Ian run a 1000 m race. Fig. E2.8 shows distance–time graphs for both runners. Sean's line is solid and Ian's line is broken.

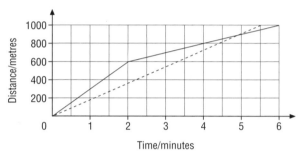

Fig. E2.8

a) Who won? By how many seconds? [2]
b) What was Sean's speed during the first two minutes? [2]
c) Imagine that you are a sports commentator. How would you describe the progress of the race? [4]

15 Fig. E2.9 shows the prices, in £, for the Hotel Inferno.

Hotel Inferno				
Rate per person arriving on date shown				
	Adult	Child	Adult	Child
	7 days		14 days	
1 Aug to 7 Aug	369	290	560	460
8 Aug to 14 Aug	379	320	570	470
15 Aug to 21 Aug	379	320	570	470
22 Aug to 28 Aug	369	290	560	460

Fig. E2.9

a) Mr Mann arrives on 12th August and stays for 7 days. What is his bill? [1]
b) The Levy family consists of two adults and three children. They arrive on 14th August and stay for 14 days. What is their bill? [3]
c) The value of the £ rises, and all prices are reduced by 5%. What is the new bill for the Levy family? [3]
d) Mrs Levy has taken £800 for other expenses. How many lire does she get at an exchange rate of 2700 lire per £? [2]

16 The students in a class raise money for a charity by various events. The amounts raised are shown in the table below.

Raffle	Sponsored run	Posters	Bazaar stall
£350	£120	£80	£170

a) What was the total amount raised? [1]
b) Draw a pie chart on Fig. E2.10 to show the amounts raised by the different events. [3]
c) Ten students took part in the sponsored run. What was the average amount raised per student? [1]
d) All the students sold tickets for the raffle. The average amount raised per student was £14. How many students are there in the class? [2]
e) The parents of the students agree to increase the amount raised by 20%. What is the new total amount? [2]

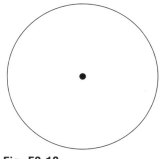

Fig. E2.10

17 Sarah wants to find out how frequently people go to watch football matches. Design a form that she could use. [4]

18 Solve the equations:

a) $x + 3 = 8$
b) $7x = 49$ [2]

19 A plumber charges a 'call-out' fee of £15 and then charges £20 per hour.

a) How much does he charge for a job that takes 3 hours? [1]
b) How much does he charge for a job that takes x hours? [2]
c) If he charges someone £95, how long has he worked for? [2]

20 A group of people took the driving test. The table below gives the numbers of people who passed or failed and their ages.

Age	17	18	19	20
Pass	10	5	7	5
Fail	8	5	8	8

a) What fraction of the people passed? [2]
b) At which age were people most likely to pass? Give a reason. [3]

21 Fig. E2.11 shows three points A, B and C on a grid. The point D is such that ABCD is a parallelogram. Mark D on the grid and write down its coordinates. [3]

[Total 100]

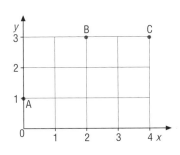

Fig. E2.11

Mock exam 3 (Intermediate)

Marks

1 A sum of money is divided in the ratio 3 : 5. What fraction of the total is the larger part? [1]

2 Ms Tanaka receives a telephone bill. Part of it is shown in Fig. E3.1, with some figures missing.

```
Your
bill is    £ _____        a   Call charges
                                £25.65  for direct dialled calls of £0.40 or under
                                £26.58  for direct dialled calls (itemised)

plus       £ _____        b   Advance charges
           ─────────
           £ _____            76.10  Subtotal excluding VAT

plus       £ _____        c   VAT at 17½%

           £ _____        d   Total amount now due
```

Fig. E3.1

   a) What are the call charges, a? [1]
   b) How much are the advance charges, b? [1]
   c) How much is the VAT, c? [1]
   d) What is the total amount now due, d? [1]

3 Fig. E3.2 is a travel graph for Mrs Barfoot's journey on a motorway. She drove at a steady speed, then stopped at a service station, then continued on her journey.

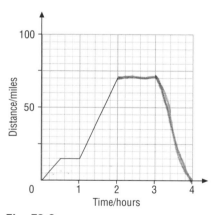

Fig. E3.2

   a) i) For how long did she stop at the service station? [1]
      ii) What was her speed during the first stage of the journey, in m.p.h.? [2]
   b) At her destination she stays for half an hour, then drives home at a steady speed of 70 m.p.h., without a break. Extend the travel graph of Fig. E3.2 to show her complete journey. [2]

4 a) Expand and simplify $(x + 3)(x - 7)$. [2]
   b) Solve the equation $x^2 - 7x + 12 = 0$. [4]

5 Fig. E3.3 shows a puzzle from a magazine. The symbols represent unknown numbers. The totals of two columns are given, and you are asked to find the total of the top row.
Let the triangle represent x and the circle represent y. Write down two equations in x and y. [2]
Solve the equations to find x and y. [2]
What is the total of the top row? [1]

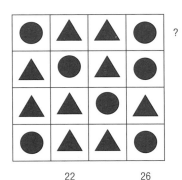
Fig. E3.3

6 a) In a sale, all shoe prices are reduced by 20%. The sale price of a pair of shoes is £48. What was the price before the reduction? [3]
b) The population of a new town is expanding at 10% each year. At the beginning of 1997 the population is 100 000. What is the population after three years have passed? [3]

7 a) Fig. E3.4 shows a pyramid with a square base. On the diagram indicate a plane of symmetry. [3]
b) On Fig. E3.5 draw the lines of symmetry of the shape. [2]
c) Complete Fig. E3.6 so that it has rotational symmetry of order 4 about the point labelled X. [2]

Fig. E3.4

8 In a bookshop, Tracy buys books costing £4.29, £7.89 and £3.50, and gives a £20 note to the assistant. Show how she can work out the approximate amount of change she should get, without using a calculator. [3]

9 Fig. E3.7 shows a pole of length 5 m leaning against a wall. The top of the pole is 1.5 m above the base of the wall.

a) Find the horizontal distance between top and bottom of the pole. [2]
b) Find the angle of slope of the pole. [2]
c) The pole slips, so that its angle of slope is 10°. Find the new height of the end of the pole above the base of the wall. [2]

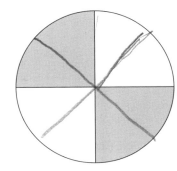
Fig. E3.5

10 a) A number is divisible by 9 if the sum of its digits is divisible by 9. What is the first number after 2145 that is divisible by 9? [2]
b) Describe briefly how you can tell whether a number is divisible by 5. [3]

Fig. E3.6

11 a) In Fig. E3.8 the triangle T has been translated to T'. Write down the vector that describes this translation. [2]

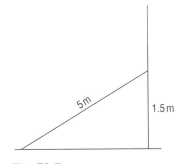
Fig. E3.7

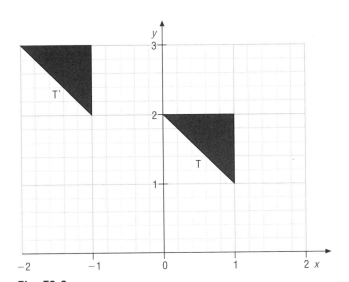
Fig. E3.8

b) The triangle T is rotated 90° anti-clockwise about (1, 1). Draw the new triangle on Fig. E3.8, and write down the coordinates of its vertices. [3]

12 Over a period of eight years the average price in pence of a standard loaf of bread was recorded. The results are in the table below.

Year	1	2	3	4	5	6	7	8
Price	40	37	42	48	48	55	56	55

Plot these values on the graph of Fig. E3.9. What sort of correlation is shown? [1]

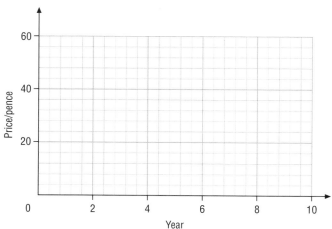

Fig. E3.9

Draw a line of best fit through the points. [2]
What will the average price be in Year 9? [2]

13 Fig. E3.10 shows a fairground game. The arrow is spun, and the score is the number in the sector that the arrow points to. Each sector has an equal angle. Find the probabilities that:

a) the score is 8 [1]
b) the score is 3. [1]

The game is played more than once. Find the probabilities that:

c) the total score with two goes is 2 [2]
d) the total score with three goes is 24. [2]

Fig. E3.10

14 Copper tubing is sold at £7.50 for a $2\frac{1}{2}$ m length.

a) What is the cost per cm? [2]
b) How many lengths can be bought for £300? [2]
c) 1 metre is about 39 inches. What is the cost of the tubing per inch? [2]

The internal diameter of the tube is 0.9 cm.

d) What is the internal volume of each length? [3]

15 In triangle ABC, AB = 6 cm, BC = 5 cm and CA = 7 cm.
See Fig. E3.11.
Make an accurate construction of △ABC. [3]
Measure $\hat{B}$ on your drawing. [2]

Fig. E3.11

16 The function y is given in terms of x by $y = x^2 - 3x + 1$. Complete the table below.

x	−1	0	1	2	3	4
y						

[3]

Plot the graph of the function on Fig. E3.12. [3]

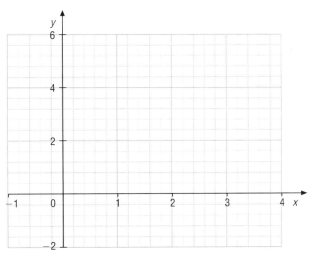

Fig. E3.12

From your graph, find the approximate solutions of the equation:

$x^2 - 3x + 1 = 0$ [2]

17 The formulae below relate to lengths, areas or volumes. The numbers 2, $\frac{2}{3}$ and π have no dimension. The letters r, h and l stand for lengths. Circle those expressions that give volumes.

$\frac{2}{3}\pi r^3$ $2\pi r(r + h)$ $r^2 h$ $\pi r l$ $2(rh + hl + lr)$ [3]

18 The cumulative frequency graph of Fig. E3.13 shows the times obtained by 400 runners of a 5-mile race.

a) Find the median time. [2]
b) Find the inter-quartile range. [2]

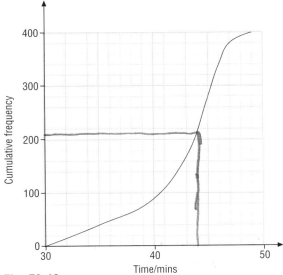

Fig. E3.13

A week later, the same runners run a 5 mile race under different conditions. The table below gives their times.

Time (mins)	30–35	35–40	40–45	45–50
Frequency	140	168	77	15

    c) Find the cumulative frequencies. [2]
    d) Plot a cumulative frequency curve on the graph of Fig. E3.13. [3]
    e) Leo ran the first race in 37 minutes. What do you expect was his time in the second race? [2]

[Total 100]

Mock exam 4 (Intermediate)

Marks

1. An author buys a computer for £999 and a printer for £349. VAT at $17\frac{1}{2}\%$ is added. What is the total cost? [2]

2. a) Solve the equation $3(x + 1) - 2(x - 3) = 8$. [2]
   b) Expand and simplify $(a + 2b)(3a + b)$. [2]
   c) Make x the subject of the formula $2y = 3x - 4$. [2]

3. A coin is bent, so that the probability that it comes up Heads is $\frac{1}{4}$. The coin is spun once.

   a) What is the probability that Heads does not come up? [1]

   The coin is spun twice.

   b) Complete the tree diagram of Fig. E4.1. [2]
   c) i) What is the probability of Heads on both spins? [2]
      ii) What is the probability that Heads comes up on exactly one of the spins? [2]

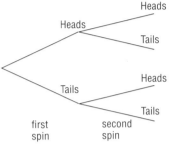

Fig. E4.1

4. Fig. E4.2 shows a heavy beam of length $2\frac{1}{2}$ m. It is held up by a vertical post.
   a) If the post is of length 1 m, what angle does the beam make with the horizontal? [2]
   b) What length of post will keep the beam at 45° to the horizontal? [2]

Fig. E4.2

5. A list of numbers is given below.

   $-5, \ -3, \ 3, \ 6, \ 7, \ 25$

   a) From the list, find a number which is:
      i) square ii) even [2]
   b) From the list, find a pair of numbers which has:
      i) sum 4 ii) product 15 [2]
   c) What is the greatest difference between numbers in the list? [1]
   d) Which pair of numbers have a difference of 10? [1]

6. a) Evaluate the expression below using a calculator. Give your answer correct to 4 significant figures.

   $$\pi \sqrt{\left(\frac{18.23}{2.15}\right)}$$ [3]

   b) How could you find the approximate value of the expression without a calculator? [3]

7. There is a positive solution to the equation $x^3 + x = 5$. Find the solution correct to 1 decimal place by trial and improvement. Use the table below.

$x = 0$	$x^3 + x = 0$	low
$x = 1$	$x^3 + x = 2$	low
$x =$	$x^3 + x =$	
$x =$	$x^3 + x =$	[4]

8   The length of a certain virus is 0.000 000 008 m.

   a) Write this number in standard form. [2]
   b) How many of these viruses, laid end to end, would stretch for 20 m? Give your answer in standard form. [2]

9   A university course involves continuous assessment, a project and a final exam. The total mark is obtained from these component marks in the ratio 3 : 2 : 5.

   a) What fraction of the total mark is obtained from the project? [2]
   b) A student got 80% in the continuous assessment, 70% in the project and 66% in the final exam. What was her total percentage mark? [2]

10 The table below gives the ages of 130 employees of a firm.

Age	20–24	25–29	30–34	35–39	40–44	45–49
Frequency	39	31	23	18	10	9

   a) Estimate the mean age of the employees. Give your answer correct to 1 decimal place. [3]
   b) What is the modal interval? [1]
   c) In which interval does the median lie? [2]
   d) On Fig. E4.3 construct a bar chart for the data. [3]

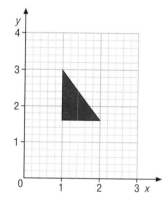

Fig. E4.3

11 a) A length is measured as 23 m, correct to the nearest metre. What is the least possible value of the length? [1]
   b) A weight is given as 0.270 kg, correct to 3 significant figures. Write down the limits between which the weight lies. [2]

12 Enlarge the triangle of Fig. E4.4 by a scale factor of $1\frac{1}{2}$, with centre of enlargement (1, 4). Write down the coordinates of the new vertices. [4]

13 Fig. E4.5 shows isometric grids.

   a) A cuboid is 2 units by 3 units by 1 unit. Draw the cuboid on the first grid. One side has already been drawn for you. [2]
   b) Place point Z on the second grid, given that the triangles ABC and XYZ are congruent. [2]

Fig. E4.4

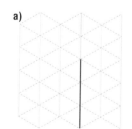 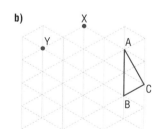

Fig. E4.5

14 Christine wants to find the height of a building. She plants a 2 m stick vertically in the ground, and walks backwards until the top of the stick appears in line with the top of the building, as shown in Fig. E4.6.

She has walked 5 m back from the stick, and the base of the stick is 50 m from the base of the building. If her eyes are 1.5 m above the ground, find the height of the building. [4]

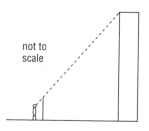

Fig. E4.6

15 Fig. E4.7 shows a map of a region of sea. A and B are lighthouses, and G is a gun emplacement. The scale of the map is 1 cm for 10 km. A ship sails so that it is an equal distance from A and B.

  a) Plot the course of the ship on the diagram. [3]
  b) The range of the gun is 20 km. On the diagram indicate the region which the gun can shell. Is the ship ever in danger? [3]

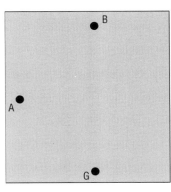

Fig. E4.7

16 Fig. E4.8 shows an isosceles triangle, in which the angles are $3x°$, $(6x + y)°$ and $(3x + 3y)°$.

  a) Write down two equations in x and y. [2]
  b) Show that these equations can be simplified to:

  $$3x - 2y = 0$$
  $$3x + y = 45$$ [2]

  c) On the graph of Fig. E4.9 draw the graphs of the equations:

  $$3x - 2y = 0$$
  $$3x + y = 45$$ [4]

  d) Find the values of x and y. [2]

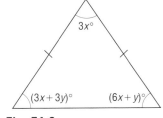

Fig. E4.8

17

Fig. E4.10

Fig. E4.10 shows a sequence of patterns made with sticks.
  a) Draw the next pattern in the sequence. [2]
  b) Fill in the table below.

Pattern number	1	2	3	4	5
Number of sticks	4				

[2]

  c) How many sticks are there in the nth pattern? [3]

18 a) Solve the inequality $3x + 1 > x + 7$. [2]
  b) Fig. E4.11 shows the graph of $y = x^2 - 2x$. Use it to solve the inequality:

  $$x^2 - 2x < 0$$ [2]

19 The diameter of a bicycle wheel is 26 inches.

  a) What is the circumference of the wheel, in inches? [2]
  b) What is the circumference of the wheel in metres, given that 1 inch is equal to 2.54 cm? [2]
  c) The bicycle is ridden for a journey of 10 km. How many times has the wheel turned? [2]

[Total 100]

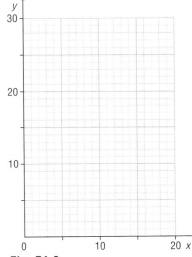

Fig. E4.9

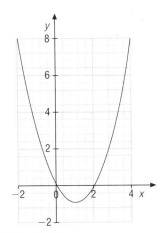

Fig. E4.11

SOLUTIONS

Solutions

Chapter 1 Operations on numbers

Exercise 1A — page 3

1. 1020 kg
2. £7
3. 1985
4. £1375
5. 58
6. 119
7. 16, 4
8. 3, 2
9. 36, 5
10. 19, 1
11. $1 \times 17 + 4 \times 20$
12. a) $A + B$ b) $3B$ c) $2A + B$
13. a) 13 b) 4 c) 25 d) 16

Exercise 1B — page 4

1. a) 11 is prime b) $22 = 2 \times 11$ c) $35 = 5 \times 7$ d) $12 = 2 \times 2 \times 3$ e) $25 = 5 \times 5$
2. 12 by 1 or 3 by 4
3. 1 by 20, 2 by 10 or 4 by 5
4. a) multiple b) square c) prime/even d) prime e) factor f) even
5. a) 28 b) 36 c) 15 d) 2
7. b)
8. b), c) true. a), d) false

Exercise 1C — page 6

1. a) 17 b) 35 c) 0 e) 36 e) 6 f) 3 g) 16 h) 6
2. £110
3. 4500
4. 84 grams
5. £32
6. 260p
7. 10 litres
8. 8
9. £2000

Chapter 2 Types of number

Exercise 2A — page 8

1. a) one hundred and eighty-four
   b) two thousand one hundred and five
   c) six million two hundred thousand
   d) sixty two million three hundred and twenty thousand
2. a) 3427 b) 513 618 c) 700 037
3. a) 3018 b) 90 c) 1000
4. a) 317 b) 3999 c) 999 999
5. 120 000 000
6. 3000
7. 8765, 5678
8. three hundred and twenty; thirty two thousand, 32 000
9. 570, five hundred and seventy
10. a) 34 b)
11. 30; 3000

Exercise 2B — page 9

1. a) $\frac{2}{3}$ b) $\frac{1}{10}$ c) $\frac{7}{10}$
2. a) a half b) two fifths c) four ninths
3. a) $\frac{1}{2}$ b) $\frac{2}{3}$ c) $\frac{3}{4}$
4. a) $1\frac{2}{5}$ b) $1\frac{4}{7}$ c) $2\frac{2}{5}$
5. a) $\frac{5}{2}$ b) $\frac{9}{8}$ c) $\frac{11}{3}$
6. a) 12 b) 10 c) 4
7. $\frac{1}{4}$
8. 640
9. $\frac{1}{4}$
10. a) $\frac{1}{4}$ b) $\frac{1}{2}$ c) $\frac{4}{9}$
11. a) 30 b) $\frac{1}{4}$
12. $\frac{1}{5}$
13. £80
14. 25 000, twenty-five thousand

Exercise 2C — page 11

1. a) 6 b) 3
2. a) $\frac{7}{10}$ b) $\frac{1}{4}$ c) $\frac{29}{100}$ d) $\frac{1}{8}$ e) $\frac{17}{25}$
3. a) 0.25 b) 0.6 c) 0.65 d) 0.875 e) 0.003
4. a) 0.333 b) 0.556 c) 0.286 d) 0.385 e) 0.167
5. a) $1\frac{1}{5}, \frac{6}{5}$ b) $2\frac{3}{4}, \frac{11}{4}$ c) $3\frac{17}{20}, \frac{77}{20}$ d) $4\frac{6}{25}, \frac{106}{25}$ e) $6\frac{7}{8}, \frac{55}{8}$
6. 2.3
7. 3.12
8. $2\frac{2}{7}, 2.29, 2.3, 2\frac{1}{3}$

Exercise 2D — page 12

1. a) 24 b) -8 c) 8 d) -4 e) -3 f) 3
2. [number line with a), c) at -1, d) at 1, b) at between 1 and 2]
3. 300 m
4. 630 miles
5. £520
6. a) 7 hours b) $+3$ hours
7. a) -2.9 m b) 5 to 6
8. $-6 < -4 < -3 < -1 < 0 < 4 < 5$
9. $1 > \frac{1}{2} > \frac{1}{4} > -0.1 > -\frac{1}{3} > -1$
10. a) 1795, 1774 b) 16, -101

Chapter 3 Calculation

Exercise 3A — page 14

1. a) 1 500 000 b) 600 000 c) 1 400 000 d) 20 000 e) 250 f) 2
2. 160 000 kg
3. 800 miles
4. £600 000
5. 40 000

Exercise 3B — page 15

1. a) 1268 b) 112.3 c) 352 d) 49.3 e) 1827 f) 3621 g) 25 h) 17
2. 3492 kg
3. £22.65
4. 17 784
5. 50 minutes

SOLUTIONS

Exercise 3C — page 17
1. a) 15.59 b) 394.4 c) 1.5
   d) 6.45 e) 20.8 f) 360
   g) 31.36 h) 3224.8 i) 24
2. Only 3.5 divides 1.23; should be 1.23 ÷ (3.5 + 2.7) =
3. 9.3 multiplies the top line; should be 6.7 + 4.5 = ÷ (2.8 × 9.3) =
4. Evaluates b)
   a) 2 + 5 = ÷ (3 + 4) =
   c) 2 + 5 ÷ (3 + 4) =
   d) 2 + 5 = ÷ 3 + 4 =

Exercise 3D — page 19
1. a) $\frac{2}{15}$ b) $\frac{12}{77}$ c) $\frac{8}{15}$
   d) $\frac{9}{40}$ e) 3 f) $\frac{9}{16}$
   g) $\frac{4}{7}$ h) $\frac{3}{40}$ i) 12
   j) $3\frac{1}{3}$ k) $7\frac{7}{20}$ l) $\frac{64}{75}$
2. a) $\frac{17}{24}$ b) $\frac{25}{63}$ c) $\frac{17}{36}$
   d) $\frac{7}{15}$ e) $\frac{4}{9}$ f) $3\frac{7}{12}$
   g) $\frac{8}{15}$ h) $-\frac{3}{8}$ i) $1\frac{17}{20}$
3. $\frac{1}{12}$
4. $\frac{1}{40}$ tonne
5. $\frac{4}{25}$
6. 4 hours
7. $\frac{7}{40}$
8. $\frac{7}{12}$
9. $1\frac{1}{12}$ m
10. $\frac{5}{24}$ gallon

Chapter 4 Percentages and ratio

Exercise 4A — page 20
1. £3500 2. £119 3. 1 000 000
4. £900 5. 3.6 litres 6. 19.2 kg
7. £318 000 8. 14.4 stones 9. £105
10. £51.70

Exercise 4B — page 22
1. a) $\frac{3}{10}$ b) $\frac{1}{4}$ c) $\frac{4}{5}$
   d) $\frac{1}{50}$
2. a) 25% b) 75% c) 5%
   d) 40%
3. a) 43% b) 82% c) 2%
4. a) 0.4 b) 0.65 c) 0.04
5. 0.2, $\frac{1}{4}$, 30%, $\frac{1}{3}$, $\frac{2}{5}$

Exercise 4C — page 22
1. 37.5% 2. 40.625% 3. 40%
4. 20% 5. 12.5% 6. 25% 7. 20%

Exercise 4D — page 23
1. £13 310 000 2. £1157.63 3. £5527.13
4. 1.458 kg 5. £18 000 6. £1500
7. £500 8. £900 9. £80
10. £4400

Exercise 4E — page 24
1. 5 : 6 2. 4 : 5 3. 1 : 4
4. 5 : 1 5. 2 : 1 6. 8 ounces
7. 180 cm 8. 27 kg 9. £25 000
10. 9 oz 11. £22.50 12. 240 miles
13. 288 grams 14. £160
15. $1\frac{1}{2}$ lb mushrooms, 9 tbsp oil, 3 lemons, 3 cloves garlic, 12 tbsp parsley
16. 625 g flour, 5 tsp cream of tartar, $2\frac{1}{2}$ tsp bicarb., 125 g butter, 75 g sugar, 125 ml milk

Exercise 4F — page 26
1. 800 2. 13.5 kg 3. £15 000
4. a) £400 000 b) £10 000 5. £75
6. £16 000, £20 000, £24 000 7. 40
8. 3.3 m, 3 m, 2.7 m

Chapter 5 Powers

Exercise 5A — page 28
1. a) 9 b) 81 c) 25
   d) 121
2. a) 5 b) 2 c) 8
   d) 12
3. a) 27 b) 125 c) 64
   d) 1000
4. a) $\frac{1}{9}$ b) $\frac{1}{8}$ c) $\frac{1}{2}$
5. 2500 m^2 6. 16 square feet
7. 80 cm 8. 13 inches 9. 125 cm^3

Exercise 5B — page 30
1. a) 10 000 b) 625 c) 1
   d) 512
2. a) $\frac{1}{4}$ b) $\frac{1}{8}$ c) 1
   d) $\frac{1}{1000}$
3. a) 2^8 b) 3^5 c) 5^5
   d) 2^3 e) 7^7 f) 8^7
4. a) 1.732 b) 0.707 c) 6.633
   d) 0.173
5. a) 3.2 b) 3.16 c) 3.16228
6. a) 4.41 b) 19.27 c) 2.44
   d) 7.31
7. 2^{20} 8. 10^{18}
9. The square root was applied to 6.1 only. The correct sequence is: 3 . 2 + 6 . 1 = √
10. She squared π × 3 instead of just 3. The correct sequence is: π × 3 x^2 =

© IT IS ILLEGAL TO PHOTOCOPY THIS PAGE

Exercise 5C — page 32

1. a) 4.1×10^5 b) 9.01×10^8 c) 2.73×10^{-3} d) 7.2×10^{-6}
2. a) 6×10^{10} b) 1.2×10^{10} c) 1.5×10^{16} d) 1.8×10^5 e) 2.1×10^{-3} f) 3×10^2 g) 4.8×10^{-14} h) 2.1×10^{-16} i) 1.3×10^{16} j) 5×10^8 k) 7.2×10^6 l) 7.5×10^7
3. 6×10^{17} kg
4. 5.9×10^{12} miles
5. 5×10^{28}
6. $\$2.4 \times 10^{12}$
7. $\$3\,000\,000\,000$, $\$3 \times 10^9$
8. 0.07 m^2
9. 1.22×10^9
10. a) 1.58×10^9 km b) 1.28×10^9 km

Chapter 6 Error and approximation

Exercise 6A — page 35

1. a) 5 b) 5 c) 44 d) 66 e) -8 f) -6 g) 1 h) 0 i) 2 j) 1
2. a) 23 847 000 kg b) 23 850 000 kg
3. a) 74 b) 11 c) 6 d) 50 e) 9 f) 467
4. £16
5. £32
6. 5p
7. 280 miles
8. 6 hours
9. £14 000

Exercise 6B — page 36

1. a) 1.24 b) 4.12 c) 4.20 d) 0.00
2. a) 54.8 b) 54.76 c) 54.759
3. a) 2390 b) 309 000 c) 0.003 81
4. a) 1.21 b) 1.2 c) 1
5. a) 8.95 b) 7.11 c) 28.67
6. 1.385 g/cm^3
7. 600 g
8. a) 2.46 b) 2.4615

Exercise 6C — page 37

1. 63.5 kg, 64.5 kg
2. 47 500 000, 48 500 000
3. 17.5°C
4. 37.35 cm, 37.45 cm
5. 47.25 ha
6. 27.325 m/s, 27.315 m/s

Exercise 6D — page 38

1. a) 12.9 $(5 + 8 = 13)$
   b) 5.9 $(3 + 3 = 6)$
   c) 4.86 $(6 - 1 = 5)$
   d) 24.58 $(70 - 40 = 30)$
   e) 34.56 $(7 \times 5 = 35)$
   f) 47.58 $(10 \times 4 = 40)$
   g) 2.25 $(6 \div 3 = 2)$
   h) 10.27 $(30 \div 3 = 10)$
   i) 4.82 $(1 \times 6 \div 2 = 3)$
   j) 120.37 $(7 + 6) \times (4 + 5) = 117$
2. 900 kg
3. 90F
4. 4000 kg
5. £18
6. 180 000 pesetas
7. £180

Exercise 6E — page 39

1. 0.1 in
2. £200
3. £10 500
4. 40 000 000 m
5. a) nearest million b) 2 d.p. c) 1 d.p. d) 1 d.p. e) nearest thousand f) nearest whole unit

Chapter 7 Practical arithmetic

Exercise 7A — page 41

1. a) £2.50 b) 9 km/litre c) £12
2. a) £90 b) £2.80
3. a) 40 litres b) 1.6 lb
4. a) after 250 minutes b) 2600 cm^3
5. £5.70
6. 22 litres
7. 12.8 km/litre
8. £113.28
9. £2.31
10. £281.25
11. left-hand bottle
12. right-hand advertisement

Exercise 7B — page 42

1. a) £21 600 b) £11 700 c) £17 056
2. a) £220.40 b) £315
3. £6.80
4. £430.65
5. 4 hours
6. shop worker
7. a) £3125 b) £23 100
8. a) £140 b) £56
9. £6170
10. £60

Exercise 7C — page 43

1. a) 0600 b) 1700 c) 2030
2. a) 8 a.m. b) 2 p.m. c) 11.30 p.m.
3. 6 hours 45 minutes
4. a) 21st March b) 13 weeks
5. a) 107 km b) 180 km
6. a) 86 minutes b) Crowhurst
7. £3055.20, £3746
8. a) $1\frac{1}{2}$ hours b) 8.00; 35 minutes

Mixed exercise 1 — page 45

1. a) -4 s b) 15 s
2. 75 min
3. 1125
4. 1018 kB
5. 44
6. $\frac{1}{3600}$

7 **a)** 140 kg   **b)** 3 : 2
8 **a)** 5048   **b)** seven thousand and thirty-three
9 **a)** 19   **b)** 33
10 **a)** 765 FF
   b) the 9 only multiplied 7×5, not $(5 \times 10) + (7 \times 5)$
   c) correct sequence is $5 \times 10 + 7 \times 5 = \times 9 =$
11 37.5%   12 30 m.p.h.
13 **a)** $\frac{1}{3}$ cm/min   **b)** 1 hour   **c)** 26 cm
14 20%, 0.22, $\frac{1}{4}$, $\frac{2}{7}$ 15 second
16 **a)** £258   **b)** £400 − £100
17 5 ft   18 £26 814   19 £1288
20 3 avocados, $1\frac{1}{2}$ pt stock, $\frac{3}{8}$ pt cream, $\frac{3}{4}$ lemon
21 **a)** 12 lb   **b)** 7 weeks   **c)** 11 st 2 lb
   d) 26th February
22 $\frac{3}{5}$
23 **a)** 8   **b)** 18   **c)** 7
24 $\frac{13}{20}$ 25 8 000 000 000   26 136 800
27 225, 235
28 **a)** 4.5×10^8 **b)** 5.2×10^8
29 **a)** −30   **b)** 2
30 **a)** 6 kg   **b)** 1.2×10^{-8} kg

■ Chapter 8 Algebraic expressions

Exercise 8A   *page 49*
1 13   2 1.5   3 35
4 20   5 27   6 616
7 300   8 54   9 29
10 66   11 405 000   12 190
13 132   14 8   15 6
16 7

Exercise 8B   *page 50*
1 £$(x + 20)$, £$(2x + 20)$
2 $(x + y)$ min   3 $(t − 10)$°C
4 $180° − x° − y°$ 5 £$10x$ 6 £$6y$
7 $100x$ m²   8 £$\frac{1}{10}W$ 9 $\frac{1}{3}m$ m.p.h.
10 $x = y + 10$ 11 $(60 + x)$ min   12 £$(40 + 10x)$
13 $3n + 4$ 14 $(5x + 7y)$p   15 £$(30x + 40y)$

Exercise 8C   *page 52*
1 **a)** 11, 13   **b)** 20, 23   **c)** 9, 7
   d) 96, 192   **e)** $\frac{1}{16}$, $\frac{1}{32}$ **f)** 11, 16
2 31, 63, $2x + 1$ 3 84, 246, $3(x − 2)$
4 **a)** $2n − 1$ **b)** $5n + 1$ **c)** n^2
   d) $n^2 + 1$

5 **a)**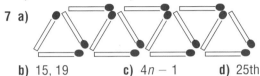
   b)
   c)

6 **a)** $2n − 1$ **b)** n^2

7 **a)**
   b) 15, 19   **c)** $4n − 1$ **d)** 25th
8 3, 5, 8

■ Chapter 9 Equations

Exercise 9A   *page 55*
1 5   2 19   3 7
4 5   5 9   6 12
7 6   8 3   9 3
10 5   11 3   12 4
13 20   14 3   15 3
16 5   17 −5   18 4
19 2   20 9   21 20

Exercise 9B   *page 56*
1 **a)** 4.1   **b)** 2.8   **c)** 3.1
  d) 4.9   **e)** 1.7   **f)** 0.6
  g) 1.8   **h)** 2.4   **i)** 2.3
2 **a)** 1.30   **b)** 0.94   **c)** 1.31

Exercise 9C   *page 58*
1 $x = 4, y = −1$ 2 $x = 2, y = 4$ 3 $x = 6, y = 1$
4 $x = 11, y = 6$ 5 $x = 2, y = 0$ 6 $p = 4, q = 3$
7 $x = 5, y = 3$ 8 $x = 7, y = 1$ 9 $x = 13, y = 7$
10 $m = 7, n = 3$ 11 $x = 4, y = 2$ 12 $x = 3, y = 1$
13 $q = 6, p = 5$ 14 $x = 4, y = 3$ 15 $x = −17, y = 14$
16 $z = 1, w = 7$ 17 $x = 5, y = 2$ 18 $q = 5, p = 5$
19 $x = 4, y = 1$ 20 $x = 3, y = −2$
21 $x = 2, y = 7$

Exercise 9D   *page 59*
1 **a)** £$(15 + 30x)$ **b)** 3 hours
2 £$(30 + 0.1x)$ **b)** 180 miles
3 **a)** $2x + 7$ **b)** 11
4 **a)** $(2x + 5)$ kg   **b)** 72 kg
5 55
6 $20x + 26(x + 2) = 604$; $x = 12$

SOLUTIONS

7 $n + 1$; $2n + 1 = 93$; $n = 46$
8 £$(x + 4)$; $38x + 7(x + 4) = 298$; $x = 6$
9 $3x + 4y = 418$, $x + 3y = 236$; 62p
10 £765, £208
11 $x + y = 1000$, $5x + 9y = 6440$; 360
12 42

Exercise 9E — page 60
1 a) $x < 3$ **b)** $x > 7$ **c)** $x \leq 3$
   d) $x < 15$ **e)** $x > 24$ **f)** $x > 6$
   g) $x \leq 5$ **h)** $x > 3$ **i)** $x \leq 4$
2 $40x + 30 < 500$; $x < 11\frac{3}{4}$
3 $10x + 20(x + 4) > 1000$; $x > 30\frac{2}{3}$

Chapter 10 Algebraic manipulation

Exercise 10A — page 63
1 $3x + 24$ **2** $5p + 5q$ **3** $12a + 4b$
4 $20x - 35y$ **5** $5x - 9$ **6** $2p + 10$
7 $11p + 7q$ **8** $-a - 12b$ **9** $x^2 + 5x + 6$
10 $z^2 + 7z + 10$ **11** $t^2 - t - 6$ **12** $x^2 - x - 30$
13 $w^2 - 11w + 30$ **14** $y^2 - 7y + 12$ **15** $x^2 - y^2$
16 $6x^2 - 5xy - 6y^2$ **17** $20a^2 - 27ab + 9b^2$
18 $x^2 + 4x + 4$ **19** $9y^2 - 12y + 4$
20 $ac + ad + bc + bd$
21 $xz - xw - yz + yw$

Exercise 10B — page 63
1 $a(b + c)$ **2** $y(x - 3)$ **3** $a(2 + b)$
4 $a(3b - 5c)$ **5** $2y(2x + z)$ **6** $2a(3b + 2c)$
7 $x(x + 3)$ **8** $y(1 + y)$ **9** $3x(x - 2)$
10 $3z(5z - 4)$ **11** $3a(3a + b)$ **12** $6p(1 + 4pq)$

Exercise 10C — page 64
1 a) $(x + 3)(x + 2)$ **b)** $(x + 2)(x + 1)$ **c)** $(x + 1)^2$
   d) $(x - 3)(x - 1)$ **e)** $(x - 4)(x - 2)$ **f)** $(x + 5)(x - 1)$
   g) $(x + 4)(x - 2)$ **h)** $(x - 6)(x + 3)$ **i)** $(x - 5)(x + 4)$
2 a) $x = -3, x = -1$ **b)** $x = -3, x = -5$
   c) $x = 5, x = 1$ **d)** $x = 5, x = 2$
   e) $x = -5, x = 1$ **f)** $x = -5, x = 3$
   g) $x = 3, x = -1$ **h)** $x = 4, x = -3$

Exercise 10D — page 65
1 a) $x = \frac{1}{3}y - \frac{2}{3}$ **b)** $m = \frac{1}{5}k + 2$ **c)** $w = \frac{1}{5}z - \frac{3}{5}$
   d) $z = w + x - y$ **e)** $x = 1\frac{1}{3} - \frac{2}{3}y$ **f)** $x = 8 - 7y$
   g) $x = 3y + 12$ **h)** $x = 4y - 12$ **i)** $z = 1\frac{1}{2}y + 12$
   j) $c = \dfrac{ab}{d}$ **k)** $x = \dfrac{y}{4a}$
   l) $a = 2b - c$ **m)** $x = \dfrac{(y - c)}{m}$
   n) $x = \dfrac{(2y + 10)}{z}$ **o)** $y = \dfrac{7}{3x}$
2 $R = \frac{4}{9}(F - 32)$ **3** $a = \dfrac{(v - u)}{t}$ **4** $h = \dfrac{V}{\pi r^2}$

Chapter 11 Coordinates and graphs

Exercise 11A — page 68
1 a) D1   **b)** A3, A4, B3, B4, C3, C4
2 (1, 3), (2, 1), (3, 2)
3

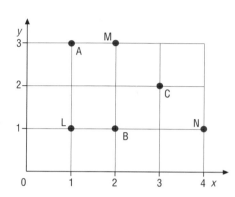

4 $(-1, 2)$, $(-1, -1)$, $(1, -2)$
5 a) $(4, 1)$ **b)** $(2\frac{1}{2}, 2)$
6 $(0.4, 1.3)$, $(1.6, 0.6)$, $(2.8, 2.5)$

7    **8**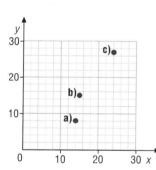

9 a) b3   **b)** d5, e5, f5, f4, f3, e3, d3, d4
   c) d2, c3, c5, d6, f6, g5, g3, f2
10 a) C6   **b)** F7   **c)** K9

Exercise 11B — page 70
1 a) 19F   **b)** £2.80
   c)

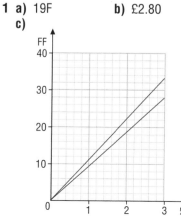

2 45DM, £5

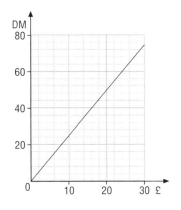

3 a) $32 **b)** £17
4 a) 11% **b)** 2 years
5 a) 104°, 96° **b)** 0400 and 1200
6 a) 24% **b)** in 1989
7 a) 4000 m, 44° **b)** 31°, 58°

8

Miles	5	10	15	20	25
Acme charge (£)	22	32	42	52	62
Nadir charge (£)	15	30	45	60	75

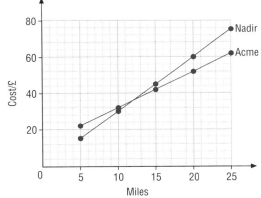

over 12 miles

9

F°	R°
0	−14
50	8
100	30

10 graph **c)**

Exercise 11C page 72

1 a) 150 m **b)** 4 min **c)** 14 m/min
2 a) 5 min **b)** 3 miles
3 a) 1 mile **b)** 15 min **c)** 56 m.p.h.
4 a) 5 seconds **b)** 9 m **c)** 0.9 m/s

5

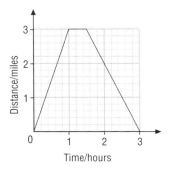

6 after 2.7 min; 130 m

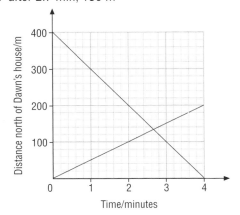

Chapter 12 Functions and graphs

Exercise 12A page 75

1

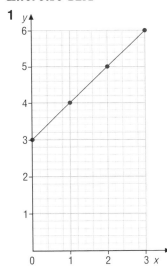

2 a)

x	0	1	2	3
y	2	3	4	5

b)

x	0	1	2	3
y	2	1	0	−1

c)

x	0	1	2	3
y	1	3	5	7

3

x	−2	−1	0	1	2
x^2	4	1	0	1	4
$x^2 + 1$	5	2	1	2	5

4

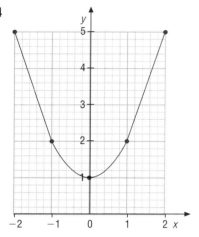

x	−2	−1	0	1	2	3
y	8	2	0	2	8	18

5

x	−1	0	$\frac{1}{2}$	1	$1\frac{1}{2}$	2	3
y	4	1	$\frac{1}{4}$	0	$\frac{1}{4}$	1	4

Exercise 12B
page 77

1

x	1	2	3	4	5
y	3	2	$1\frac{2}{3}$	$1\frac{1}{2}$	$1\frac{2}{5}$

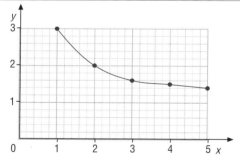

2

x	1	2	3	4
y	1	2	$2\frac{1}{3}$	$2\frac{1}{2}$

3

x	−1	0	1	2	3
x^2	1	0	1	4	9
$2x$	−2	0	2	4	6
y	3	0	−1	0	3

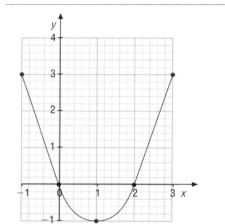

4

x	−2	−1	0	1	2	3
y	4	0	−2	−2	0	4

5 $0 < x < 2$ **6** $-1 < x < 2$

Exercise 12C
page 79

1 $x = 1, y = 3$

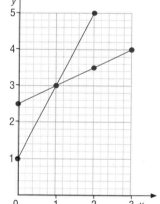

2 a) $x = 3, y = 4$ **b)** $x = 3, y = 2$
c) $x = 2, y = 3$

3

x	−4	−3	−2	−1	0	1	2
y	3	−1	−3	−3	−1	3	9

$x = -3.3$ or $x = 0.3$

4

x	½	1	1½	2	2½	3	3½	4
y	4½	3	2⁵⁄₆	3	3³⁄₁₀	3²⁄₃	4¹⁄₁₄	4½

$x = 0.6$ or $x = 3.4$

5 a) after 1.5 sec **b)** 13 m
   c) after 0.2 and 2.8 sec **d)** 3.1 sec

6 a) 145 feet **b)** 31 m.p.h.

Exercise 12D page 80

1 $x > 2, y > 1, x + y < 4$

2 A: $x > 0, 3x + 2y < 6, 2x + 3y > 6$
   B: $y > 0, 3x + 2y > 6, 2x + 3y < 6$

3
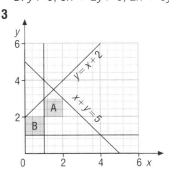

4 $x + y \leq 5, 60x + 40y \leq 240$
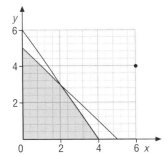

■ Mixed exercise 2 page 82

1 a) $x = 15$ **b)** $x = 4$ **c)** $x = 7$

2 8

3 a) (1.9, 2.2) **b)**
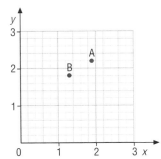

4 a)
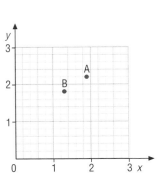

b) 4, 8, 12, 16 **c)** 20

5
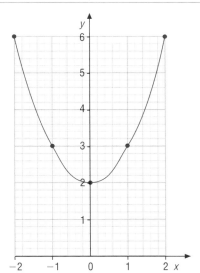

6 $5x + 13$

7 a) 240 yards **b)** 6 min **c)** 40 yards/min
   d)
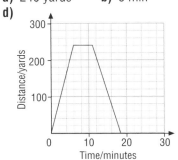

8 a) £$5x$ **b** £$(5x - 7)$ **c)** $5x - 7 = 78$
   d) $x = 17$

9 a)
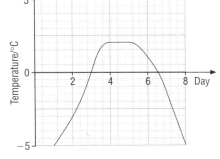

b) 7°C **c)** 7 and 8

10 a) $y = 3\frac{1}{2}$ **b)** $z = 49$

11 a) $x = 7, y = 4$ **b)** $x = 3, y = 2$

12 $4n$

13 a) i) $2x(2a + 7b)$ **ii)** $(x + 2)(x + 5)$ **b)** $x = 2$ or $x = 7$

14 a) $x = 2$ or $x = 3$
   b) $2 < x < 3$
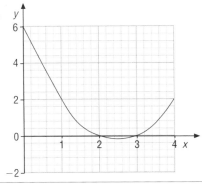

15 9.90

16 a) $v = \frac{1}{5}s - \frac{3}{5}$ **b)** $v = \frac{T}{5a}$

17 $x = 1.6$ **18** $-40°$

19 a) $x > 6$ **b)** $x > 3$

20 a) (iii)   **b)** (iv)   **c)** (i)   **d)** (ii)

Chapter 13 Plane figures

Exercise 13A   page 86
1 $a° = 130°; b° = 80°$

2 a) 180°   **b)** 90°   **c)** 720°

3 a) 2   **b)** 5   **c)** $1\frac{1}{2}$

4 a) 72°   **b)** 1.2°

5 60°   **6** 15 minutes

7 a) 90°   **b)** 180°   **c)** 150°

8 $x° = z° = 145°, y° = 35°$

9 a) acute   **b)** right
c) obtuse   **d)** reflex

Exercise 13B   page 88
1 $a° = 110°, b° = c° = 70°$ **2** 65°

3 $x° + y° = 180°$

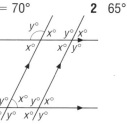

4 diagram **a)**

Exercise 13C   page 90
1 $a° = 63°, b° = 150°, c° = 74°, d° = 99°$

2 $x° = 50°, y° = 80°$

3 70°   **4** both 61°

5 both 60°, equilateral

7 right-angled, isosceles

8 a) trapezium   **b)** kite   **c)** rhombus
d) rectangle

Exercise 13D   page 91
1 a) 720°   **b)** 900°   **c)** 1440°

2 a) 108°   **b)** 162°   **c)** 140°

3 a) 60°   **b)** 30°   **c)** 20°

4 pentagon   **5** hexagon   **6** 18

7 540°; $x = 98°$

8 a) yes, 36   **b)** yes, 18   **c)** no

9 108°, 36°, 72°; isosceles

10 120°, 30°, 90°; rectangle

11 triangles, squares, hexagons

12 135°, 8

Exercise 13E   page 93
3 90°, 55°   **4** 37°   **5** $a° = 88°, b° = 45°$

Chapter 14 Measures, lengths and areas

Exercise 14A   page 95
1 a) 500 cm   **b)** 1.25 kg   **c)** 3 ha
d) 223 ml

2 a) 159 ft   **b)** 5 stones   **c)** 2640 yds
d) $12\frac{1}{2}$ gallons

3 a) 63.5 cm   **b)** 54.48 l   **c)** 52.9 lb
d) 69.2 acres

4 a) 23 km   **b)** 16 miles

5 a) 88.9 mm   **6** 216 in, 5.49 m

Exercise 14B   page 97
1 20 m.p.h.

2 a) 130 miles   **b)** $2\frac{1}{2}$ hours   **c)** 52 m.p.h.

3 a) 24 km   **b)** $2\frac{1}{2}$ hours

4 1080 kg/m³

5 a) 450 g   **b)** 6 cm³

Exercise 14C   page 97
1 120 cm   **2** 120 m, 377 m   **3** 11 cm

4 11.3 cm   **5** 14 m   **6** 159 m

7 7.16 cm   **8** 12 cm

Exercise 14D   page 98
1 a) 2 cm²   **b)** 3 cm²   **c)** 6 cm²

2 £78.75

3 a) 1200 m²   **b)** 140 m   **c)** £144
d) £1820

4 80 m   **5** 10 cm   **6** 100 m²

7 4 m²

8 a) 452 m²   **b)** 154 cm²

Exercise 14E   page 100
1 a) 20 m²   **b)** 20 m²   **c)** 16 m²

2 a) 3 cm²   **b)** 4 cm²   **c)** 3.57 cm²

3 308 m, 6427 m²   **4** 10.5 m²   **5** 100 m²

Chapter 15 Solids and volumes

Exercise 15A   page 102
1 5, 8, 5

2 a) 6, 12, 8
b) ABCD and EFGH (also others)
c) AB and EF (also others)

3 a) cylinder   **b)** cuboid   **c)** sphere
d) prism

4 a) 1   **b)** 6   **c)** 12
d) 8
5 a) 2

Exercise 15B   page 103
5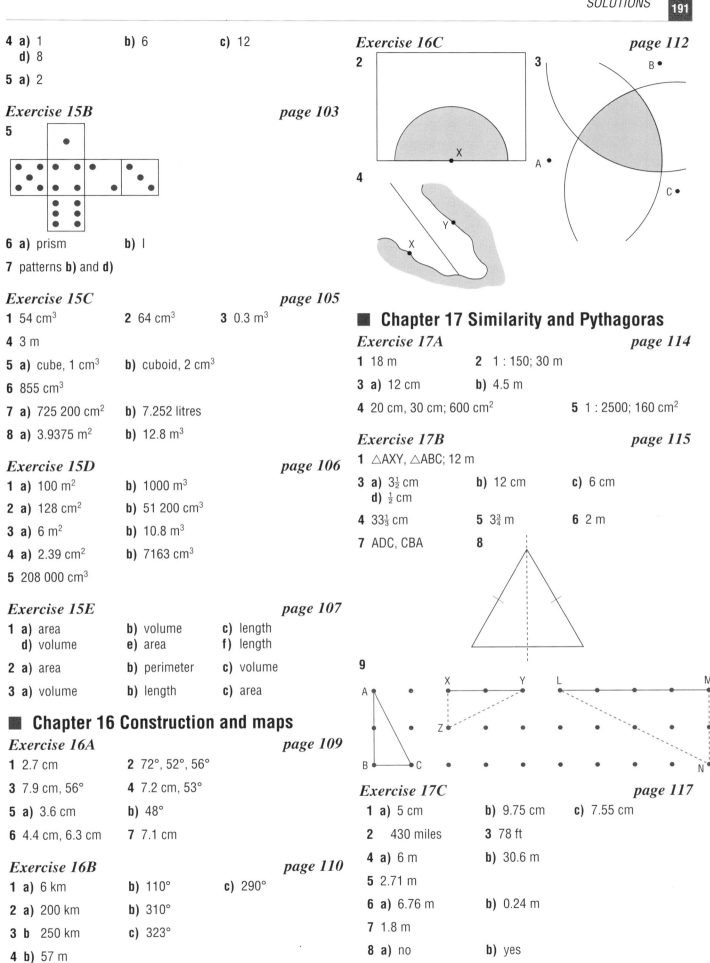

6 a) prism   **b)** l

7 patterns **b)** and **d)**

Exercise 15C   page 105
1 54 cm³   **2** 64 cm³   **3** 0.3 m³
4 3 m
5 a) cube, 1 cm³   **b)** cuboid, 2 cm³
6 855 cm³
7 a) 725 200 cm²   **b)** 7.252 litres
8 a) 3.9375 m²   **b)** 12.8 m³

Exercise 15D   page 106
1 a) 100 m²   **b)** 1000 m³
2 a) 128 cm²   **b)** 51 200 cm³
3 a) 6 m²   **b)** 10.8 m³
4 a) 2.39 cm²   **b)** 7163 cm³
5 208 000 cm³

Exercise 15E   page 107
1 a) area   **b)** volume   **c)** length
   d) volume   **e)** area   **f)** length
2 a) area   **b)** perimeter   **c)** volume
3 a) volume   **b)** length   **c)** area

Chapter 16 Construction and maps
Exercise 16A   page 109
1 2.7 cm   **2** 72°, 52°, 56°
3 7.9 cm, 56°   **4** 7.2 cm, 53°
5 a) 3.6 cm   **b)** 48°
6 4.4 cm, 6.3 cm   **7** 7.1 cm

Exercise 16B   page 110
1 a) 6 km   **b)** 110°   **c)** 290°
2 a) 200 km   **b)** 310°
3 b 250 km   **c)** 323°
4 b) 57 m
5 b) 340 m

Exercise 16C   page 112

Chapter 17 Similarity and Pythagoras
Exercise 17A   page 114
1 18 m   **2** 1 : 150; 30 m
3 a) 12 cm   **b)** 4.5 m
4 20 cm, 30 cm; 600 cm²   **5** 1 : 2500; 160 cm²

Exercise 17B   page 115
1 △AXY, △ABC; 12 m

3 a) $3\frac{1}{2}$ cm   **b)** 12 cm   **c)** 6 cm
d) $\frac{1}{2}$ cm
4 $33\frac{1}{3}$ cm   **5** $3\frac{3}{4}$ m   **6** 2 m
7 ADC, CBA   **8**

9

Exercise 17C   page 117
1 a) 5 cm   **b)** 9.75 cm   **c)** 7.55 cm
2 430 miles   **3** 78 ft
4 a) 6 m   **b)** 30.6 m
5 2.71 m
6 a) 6.76 m   **b)** 0.24 m
7 1.8 m
8 a) no   **b)** yes
9 3.61 m   **10** 15.9 cm

Chapter 18 Trigonometry

Exercise 18A — page 121

1. a) 7.00 cm   b) 6.11 cm   c) 12.6 cm
2. 4.77 cm, 7.63 cm
3. 34.4 m   4. 1.06 m   5. 37.4 m
6. a) 600 m   b) 544 m
7. 66.4 m
8. a) 124 km   b) 638 km
9. 10.5 m   10. 14.3 ft
11. 441 m, 235 m, 51 800 m²

Exercise 18B — page 123

1. a) 58°   b) 40°   c) 33°
2. 49°, 41°   3. 43°   4. 21°
5. 31°   6. 2.9°   7. 051°

Chapter 19 Transformations

Exercise 19A — page 125

1.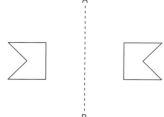

2.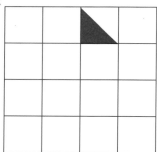

3. (1.5, 3), (2, 2.5), (1.5, 2.5)

4. $\begin{pmatrix} 1 \\ -4 \end{pmatrix}$; (5, −2)

5.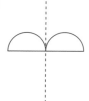

6. 4 to right   7. (1, 0), (1, 1), (3, 1)

Exercise 19B — page 126

1.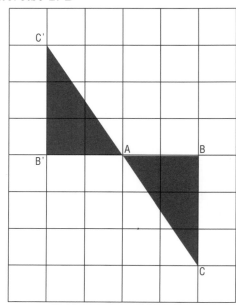

2.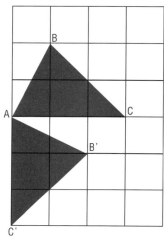

3. 90° clockwise

4.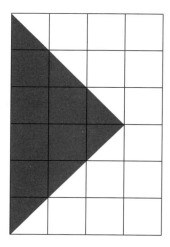

5. 2

SOLUTIONS

Exercise 19C
page 128

1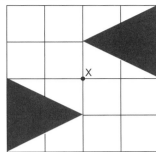

2 (0, 1), (0, 2), (−2, 2)

3 (2, 2), (4, 2), (2, 1)

4 (1, 0.4), (1, 2), (1.8, 2)

5 3, (1.5, 2)

7 90° clockwise, (0, 0)

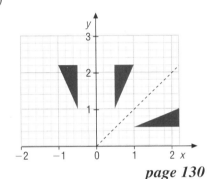

Exercise 19D
page 130

1 a) 4   b) 2   c) 1
   d) 2

2 a) 2 lines, rotational symmetry order 2
   b) 1 line of symmetry   c) no symmetry

3 5, 5

5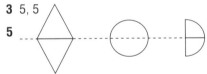

6 4   7 9

Mixed exercise 3
page 131

1 75°

2 a) 360 m   b) 0.4 m/s   c) 390 yards

3 a) 8   b) 54 cm^3

4 44°

5 a) rectangle   b) kite
   c) rhombus   d) trapezium

6 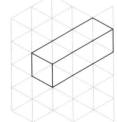   7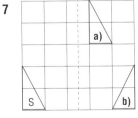

8 a) 440 m   b) 15 400 m^2

9 a) hexagon   b) 6
   c) i) 120°   ii) X

10 a) 1 : 200   b) 10 m

11 A and E, C and F   12 7 cm^2, 16 cm

14 a) 76 m   b) 18°

15 a) 2   b) (0, 1.5)

16 a) 3420 cm^3 b) 14.9 cm

17 a) 90°, 51°   b) i) equal   ii) congruent

18 AXB, DXC; 9 cm

19 a) The unit of volume must be cubed. This formula only gives $\pi r^2 h + \pi r^2$
   b) $\pi r(rh + r^2)$ (many others)

20 a) 4   b) 1

21 117°   22 c) yes, 36 km

Chapter 20 Collecting data

Exercise 20A
page 134

1 a) Kanemochi   b) Binbo   c) Anzen, Kanemochi

2 a) Jonathon   b) French

5 a) 13   b) 102

6 a) 140   b) 529   c) 305

7 a) 2 years old   b) 0.409

Exercise 20C
page 136

1 8, 10, 3, 3, 4   2 8, 6, 13, 3, 2, 14

3 5, 10, 9, 6, 6, 4   4 9, 8, 6, 5, 2

5 B: 6, C: 13, D: 4, X: 2

6 red: 3, blue: 5, black: 5, green: 1, white: 6, yellow: 1

Chapter 21 Pictures of data

Exercise 21A
page 141

1 3500

2 350, 225, 300, 250, 425, 275

3 a) 13   b) 34

4 a) 15   b) 30°

5 a) 72°   b) 11

Exercise 21B
page 143

1 117°, 45°, 27°, 171°   2 92°, 72°, 152°, 44°

3 36°, 108°, 126°, 90°

Exercise 21D
page 146

1 negative   2 positive   3 $4\frac{1}{2}$ years

4 £28 000   5 59%

Chapter 22 Analysing data

Exercise 22A
page 148

1 203.7, 208   2 1.42, 1   3 28°C, 11°C

4 131.6, 128$\frac{1}{2}$, 130   5 61.8, 63, 40   6 6.22, 7

SOLUTIONS

Exercise 22B   page 150
1. 2.16, 2
2. 0, 2.325, 2
3. 65–70, 67.9°F
4. £150–200, £155.5

Exercise 22C   page 152
1. a) 850   b) 64   c) 20
2. a)

| Cumulative frequency | 5 | 22 | 47 | 77 | 90 | 100 |

b)

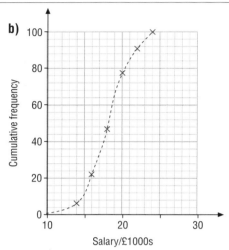

c) £18 200, £16 500, £19 900, £3400
d) 0.16

3. a)

| Cumulative frequency | 13 | 48 | 74 | 80 |

c) 29 kg, 23 kg, 34 kg, 11 kg

4. a)

| Cumulative frequency Resort A | 2 | 10 | 28 | 47 | 57 | 60 |
| Cumulative frequency Resort B | 6 | 18 | 31 | 43 | 52 | 60 |

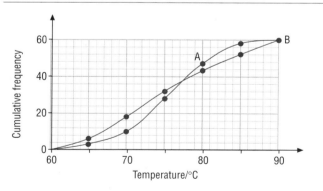

b) A: 8°, B: 12°

5. a)

| Cumulative frequency | 121 | 422 | 750 | 930 | 1000 |

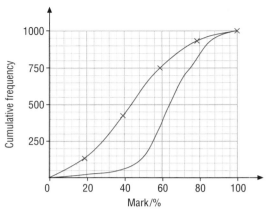

b) 50

Chapter 23 Probability

Exercise 23A   page 155
3. $\frac{1}{2}$ 4. $\frac{1}{6}$ 5. $\frac{1}{8}$
6. $\frac{7}{12}$ 7. $\frac{3}{7}$
8. a) $\frac{49}{50}$ b) 5
9. a) $\frac{2}{25}$ b) 30
10. a) $\frac{1}{6}$ b) $\frac{1}{3}$ c) 0
11. a) $\frac{1}{10}$ b) $\frac{2}{5}$
12. a) $\frac{1}{13}$ b) $\frac{1}{52}$ c) $\frac{12}{13}$
13. $\frac{5}{7}$ 14. $\frac{69}{70}$

Exercise 23B   page 156
1. AC AD AE BC BD BE; $\frac{1}{6}$
2. RB RG RW BB BG BW PB PG PW; $\frac{1}{9}$
3. a) $\frac{1}{16}$ b) $\frac{1}{4}$ c) 0
4. a)

		Score on first die					
		1	2	3	4	5	6
Score on second die	1	2	3	4	5	6	7
	2	3	4	5	6	7	8
	3	4	5	6	7	8	9
	4	5	6	7	8	9	10
	5	6	7	8	9	10	11
	6	7	8	9	10	11	12

b) i) $\frac{1}{36}$ ii) $\frac{5}{36}$ c) 7, $\frac{1}{6}$

5. HH HT TH TT; $\frac{1}{4}$
6. ABC ACB BAC BCA CAB CBA; $\frac{1}{6}$

Exercise 23C   page 158
1. $\frac{1}{78}$ 2. $\frac{1}{4}$
3. a) $\frac{1}{20}$ b) $\frac{3}{5}$

4 a) $\frac{1}{9}$ **b)** $\frac{4}{9}$

5 a) $\frac{1}{400}$ **b)** $\frac{361}{400}$

6 a) $\frac{6}{35}$ **b** $\frac{17}{35}$

7 a) $\frac{1}{100}$ **b)** $\frac{9}{50}$

8 a) $\frac{1}{64}$ **b)** $\frac{7}{32}$

9 a) $\frac{1}{2}$ **b)** $\frac{1}{6}$

■ Mixed exercise 4   *page 160*

1 $\frac{5}{12}$

3 a) negative   **c)** £80 000

4 a) 64 800   **b)** 60°

5 a) 40, 30, 25

6 60.8°, 6°

7 a) $\frac{7}{20}$ **b)** 5

8 $20\frac{1}{2}$ hours

9 a) $\frac{1}{3}$ **b)** AA AB AC BA BB BC CA CB CC
c) $\frac{1}{9}$

10 a) frequencies 9, 9, 10, 8

11 a) 110 g   **b)** 110 to 120

12 b) $\frac{3}{32}$ **c)** $\frac{25}{32}$

13 a) £61   **b)** £22
c) 14, 38, 59, 78, 100
d)

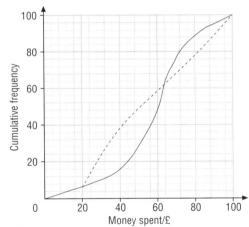

e) £48

14 a) 0.4   **b)** 32

15 a) $\frac{1}{16}$ **b)** $\frac{3}{16}$

■ Mock exam 1   *page 164*

1 four hundred and seventy-six

3 a) 0958   **b)** 11 mins

4 15 ounces

5 a) 5   **b)** 4

6 a) 900 m²   **b)** 1.7 cm³

7 £14.80, £16.28   **8** equilateral, 60°

9 a) 72 km   **b)** 160°

10 a) $\frac{1}{4}$ **b)** $\frac{2}{3}$ **c)** 0

11 a) 9, 11   **b)** 25   **c)** 1, 4, 9, 16, 25
d) square

12 a) −4°C   **b)** 7°C   **c)** 5°C

13 3   **14** TF TT TH CT CF CH; 9

16 a) totals 9, 11, 10, 6
c) 90°, 110°, 100°, 60°
d) none

17 a) i) 3.6SF   **ii)** £11
b)

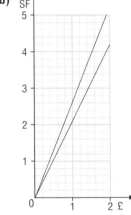

18 a) 75   **b)** 2

19 a) 2058p   **b)** 50 × 40 = 2000

20 a) $\frac{11}{40}$ **b)** £360 000

21 a) 22 miles   **b)** 59 miles   **c)** $\frac{1}{2}$ hour

22 3

■ Mock exam 2   *page 168*

1 2036   **2** £13.69; £6.31

3 a) £6   **b)** $\frac{4}{15}$

4 £65.70

5 a)

b) 12   **c)** $x + 2$

6 a) 66 inches   **b)** 167.6 cm   **c)** 1.676 m

7 a) (ii)   **b)** (iii)   **c)** (i)

8 a) 377 m   **b)** 11 300 m²   **c)** 27 times

9 a) 1   **b)** 4

10 a) all numbers <23
b) $15\frac{1}{2}$

11

x	1	2	3	4	5
y	1	3	5	7	9

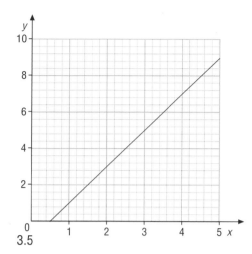
3.5

12 a) 25 min   **b)** Weather, Environment, Comedy   **c)** 75%

13 500, 250, 375; $\frac{2}{9}$

14 a) Ian, by $\frac{1}{2}$ min   **b)** 300 m/min

15 a) £379   **b)** £2550   **c)** £2422.50
    d) 2 160 000 lire

16 a) £720   **b)**
    c) £12
    d) 25
    e) £864

18 a) $x = 5$
    b) $x = 7$

19 a) £75
    b) £(15 + 20x)   **c)** 4 hours

20 a) $\frac{27}{56}$ **b)** 17 (pass rate > $\frac{1}{2}$)

21 (2, 1)

■ Mock exam 3   *page 172*

1 $\frac{5}{8}$

2 a) £52.23   **b)** £23.87   **c)** £13.32
    d) £89.42

3 a) i) $\frac{1}{2}$ hour   **ii)** 30 m.p.h.
    b)

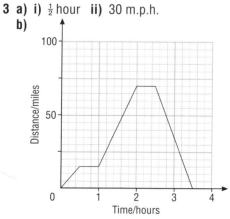

4 a) $x^2 - 4x - 21$ **b)** $x = 3$ or $x = 4$

5 $3x + y = 22$, $x + 3y = 26$; $x = 5$, $y = 7$; 24

6 a) £60   **b)** 133 100

7 a)   **b)**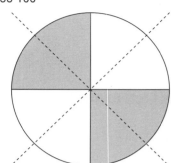

    c)

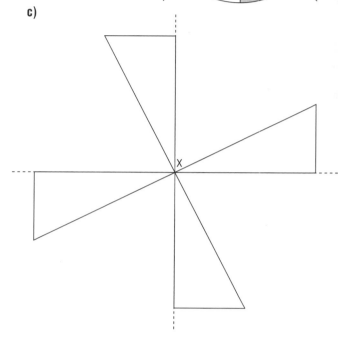

8 $20 - 4 - 8 - 4 = £4$

9 a) 4.77 m   **b)** 17°   **c)** 0.868 m

10 a) 2151   **b)** ends in 0 or 5

11 a) $\begin{pmatrix} -2 \\ 1 \end{pmatrix}$ **b)** (1, 1), (0, 1), (0, 0)

12 positive; 60p

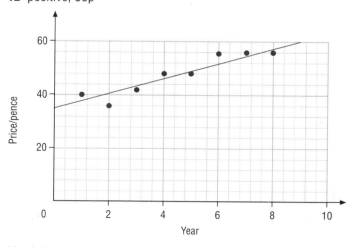

13 a) $\frac{1}{8}$ **b)** $\frac{3}{8}$ **c)** $\frac{1}{64}$ **d)** $\frac{1}{512}$

14 a) 3p b) 40 c) 7.7p d) 159 cm³
15 78.5°
16

x	−1	0	1	2	3	4
y	5	1	−1	−1	1	5

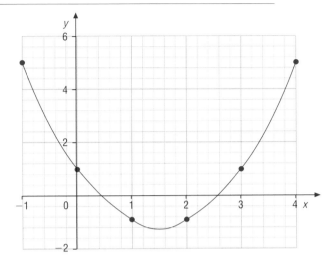

$x = 0.4$ or $x = 2.6$

17 $\frac{2}{3}\pi r^3$, $r^2 h$

18 a) 43.5 min   b) 5 min   c) 140, 308, 385, 400
d)
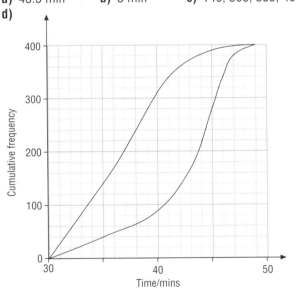
e) 32 min

Mock exam 4   page 177

1 £1583.90
2 a) $x = -1$ b) $3a^2 + 7ab + 2b^2$
   c) $x = \frac{2}{3}y + 1\frac{1}{3}$
3 a) $\frac{3}{4}$ c) i) $\frac{1}{16}$ ii) $\frac{3}{8}$
4 a) 24°   b) 1.77 m
5 a) i) 25 ii) 6   b) i) −3, 7 ii) −5, −3
   c) 30   d) −3, 7
6 a) 9.148   b) $3 \times \sqrt{(18/2)} = 3 \times 3 = 9$
7 1.5
8 a) 8×10^{-9} m   b) 2.5×10^9
9 a) $\frac{1}{5}$ b) 71%
10 a) 30.3   b) 20–24   c) 25–29
11 a) 22.5 m   b) 0.2695 to 0.2705
12 (1, 2.5), (1, 0.4), (2.5, 0.4)
13 a) 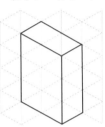   b)

14 7 m
15 b) yes
16 a) $3x + 6x + y + 3x + 3y = 180$, $3x + 3y = 6x + y$
   d) $x = 10$, $y = 15$
   c)
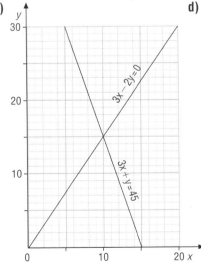

17 a)
b) 4, 7, 10, 13, 16
c) $3n + 1$
18 a) $x > 3$ b) $0 < x < 2$
19 a) 81.7 in   b) 2.07 m   c) 4820

Glossary

acute angle Angle less than 90°
acute-angled triangle Triangle with all angles less than 90°
arc Length of circle between two points

chord Straight line joining two points on a circle
composite number Number with divisors apart from 1 and itself
cone Solid which tapers to a point from a circle
cube Solid with six square faces
cuboid Solid with six rectangular faces
cumulative frequency The number of items in a set of numbers which are less than a certain value
cylinder Solid with constant circular cross-section

denominator The bottom number of a fraction
diameter Line across a circle through its centre

equilateral triangle Triangle with all sides equal
even number Number divisible by 2

hemisphere Half a sphere
hexagon Six-sided figure

interquartile range The difference between the lower quartile and the upper quartile
isosceles triangle Triangle with two sides equal

kite Quadrilateral with two pairs of adjacent sides equal

lower quartile The value in a set of numbers which cuts off the lower quarter

mean The average of n numbers, found by adding them and dividing by n
median The average of numbers found by arranging them in order and taking the middle number
mixed number Number with a whole number part and a fraction part
mode The most common number of a set of numbers

numerator The top number of a fraction

obtuse angle Angle between 90° and 180°
obtuse-angled triangle Triangle with one angle greater than 90°
odd number Number not divisible by 2

parallel In the same direction
parallelogram Quadrilateral with two pairs of opposite sides parallel
pentagon Five-sided figure
perimeter The distance around a shape
perpendicular At right angles
polygon Many-sided figure
prime number Number with no divisors, apart from 1 and itself
prism Solid with constant cross-section
pyramid Solid which tapers to a point, usually from a square base

quadrilateral Four-sided figure

rectangle Quadrilateral with each angle equal to 90°
range The difference between the highest and lowest values of a set of numbers
reflex angle Angle greater than 180°
regular polygon Polygon with all sides equal and all angles equal
rhombus Quadrilateral with all four sides equal
right angle Angle of 90°
right-angled triangle Triangle with one angle of 90°

scalene triangle Triangle with unequal sides
semi-circle Half a circle

tessellation Covering a plane with shapes
trapezium Quadrilateral with one pair of parallel sides
triangle Three-sided figure

upper quartile The value in a set of numbers which cuts off the top quarter

Index

accuracy, appropriate 38–9
acre 94
addition 2
   standard form 31
adjacent side 119
algebraic expressions 48–53
algebraic manipulation 61–6
angles 86–8
   acute 86
   alternate 87
   corresponding 87
   depression 120
   of elevation 120
   exterior 91
   interior 91
   obtuse 86
   opposite 87
   of a polygon 91
   reflex 86
   right 86
   in semi-circle 93
approximation 34–9
   checking 38
arc 92
area 98–9
   circle 98
   dimension 106
   parallelogram 98
   rectangle 98
   square 98
   trapezium 98
   triangle 98
   units 94–5
arithmetic
   of fractions 18–19
   mental 14
   on paper 15
averages 40–1, 147–50
axes 67

bar chart 140–1, 143–4
base 10 (denary) 7
bearings 109
best fit, line 145
brackets 5, 16, 49, 61, 62

calculation 14–19
calculator
   for fractions 19
   for powers 29
   scientific 16, 19, 29
   square root 27–8
   for trigonometry 119, 122
   use of 16–17
centimetre 94
centre
   of a circle 92
   of enlargement 127
   of rotation 127
changing subject 65
charts 43–4
checking, approximation 38
chord 92
circle 92–3
   area 98
circumference 92
clock, 24 hour 43
compasses 108
compound interest 23

cone 101
congruent, triangles 114–15
constructions 108–9
coordinates 67
correlation 145
cosine (cos) 119
cross-section 101
cube 101
   volume 104
cube number 27–8
cuboid 101
   volume 104
cumulative frequency 150–2
cylinder 101
   volume 104

data
   analysis 147–53
   collection 134–9
   statistical diagrams 140–6
decimal places 10, 36
decimals 10–11
   recurring 10
   terminating 10
denary 7
denominator 8
density 96
diameter 92
difference 2
dimensions 106
division 2
   in ratio 25–6
   standard form 31

edge 101
enlargement 125–8
equations 54–60
   of form $x^2 + ax + b = 0$ 64, 78–9
   in problem solving 58
   simultaneous 56–8, 78
   solving 54, 64, 78–9
error 34–9
even number 4
expansion 61–2
expressions, algebraic 48–53

face 101
factor tree 4
factorisation 63–4
factors 4
foot 94
formulae
   changing subject of 65
   making 50
fractions 8–9
   arithmetic of 18–19
   using a calculator 19
   improper 8
   proper 8
   simplification 8
frequencies 137–8
frequency polygons 143–4
frequency tables 143–4, 149–50
function 74

gallon 95
geometrical drawing 108
gram 94

graphs 69–70, 74–81
   interpretation 69–70
   linear 74–5
   reciprocal and quadratic 76–7
   for solving equations 78–9
greatest and least values 37

hectare 94
hexagon 88
hire-purchase 42
hypotenuse 116, 119

Imperial system 94
inch 94
income tax 42
independent events 157
indices 29–30
inequalities 59–60
   quadratic 76
   two-dimensional 80
inter-quartile range 150
inverse key, on calculator 16
inverse trigonometric functions 122
isometric paper 102

kilogram 94
kilometre 94
kite 89

length 97
   dimension 106
   units 94
like terms 62
line of best fit 145
line of symmetry 129
linear graph 74–5
lines 86
   parallel 86
litre 95
locus 111–12
lower quartile 150

maps 109–11
mean 147–9
measure, mixed 96
median 147, 149
memory on calculator 16
mental arithmetic 14
metre 94
metric system 94
mile 94
milligram 94
millilitre 94
millimetre 94
mixed measure 96
mixed number 8
modal group 149
mode 147
money calculations 42
multiple 4
multiplication 2
   powers 29
   standard form 31

negative correlation 145
negative numbers 11–12, 49, 59
negative powers 29

INDEX

net 102
*n*th power 29
number
  mixed 8
  operations 2–6
  types 7–13
  whole 7
number line 11
number sequence 51
numbers
  negative 11–12, 49, 59
  positive 11–12
numerator 8

odd number 4
ogive 150
operations 2–6
  order 5, 29
opposite side 119
order of operations 5, 29
order of rotational symmetry 129
origin 67
ounce 94
overtime 42

π (pi) 97
parallel lines 86
parallelogram 89
  area 98
pentagon 88
percentages 20–3
  reverse 23
perimeter 97
perpendicular lines 86
pictogram 140–1
pie chart 140–3
pint 95
plane figures 86–93
plane of symmetry 129
polygons 88–91
  angles of 91
  labelling 90
  regular 89
positive correlation 145
positive numbers 11–12
pound 94
powers 27–33
  calculator 29
  multiplication 29
prime number 4
prism 101
  volume 104
probability 154–9
problem solving, with equations 58
product 2
protractor 108
pyramid 101
  volume 104
Pythagoras' theorem 116–17

quadratic equations 78–9
quadratic function and graph 76–7
quadrilaterals 88, 89
quart 95
quartile 150
questionnaires 136

radius 92
range 147
rates 40–1
ratio 24–6
reciprocal function and graph 76–7
rectangle 89
recurring decimals 10
reflection 124
reverse percentage 23
rhombus 89
rotation 125–7
rotational symmetry 129
rounding 34–5
ruler 108

scale, of map 109
scale diagrams 113
scale factor of enlargement 125
scatter diagrams 145
sector 92
segment 92
semi-circle 92
sequences 51–2
shape sequence 51
significant figures 36
similar triangles 114–15
similarity 113–15
simplification 62–3
simultaneous, equations 56–8, 78
sine (sin) 119
solids 101–3
speed 96
sphere 101
square 89
  area 98
square number 4, 27–8
square root 27–8
  on a calculator 27–8
standard form 30
statistical diagrams 140–6
stone 94
subject of formula 65
substitution 48
subtraction 2
  standard form 31
sum 2
symmetry 129–30
  line of 129
  order of rotational 129
  plane of 129

tables 43–4, 134
tallies 137–8
tangent (tan) 119
tangent to a circle 92
terminating decimals 10
tessellation 89
tetrahedron 101
ton 94
transformations 124–30
translation 124
transversal 87
trapezium 89
  area 98
travel graphs 72
tree diagram 157–8
trial and improvement 55
triangles 88, 89
  area 98
  congruent 114–15
  equilateral 89
  isosceles 89
  right-angled 89, 116
  scalene 89
  similar 114–15
triangular prism 101
trigonometric ratios 119
trigonometry 119–23
  use of calculators 119, 122

units 94–6
unknown 54
upper quartile 150

VAT 42
vector 124
vertex 101
volume 104
  cube 104
  cuboid 104
  cylinder 104
  dimension 106
  prism 104
  pyramid 104
  units 95

weight, units 94

yard 94

zero power 29
zeros 7